INTEGRATED COMPUTER-AIDED
DESIGN OF MECHANICAL SYSTEMS

INTEGRATED COMPUTER-AIDED DESIGN OF MECHANICAL SYSTEMS

S. A. MEGUID

*School of Mechanical Engineering,
Cranfield Institute of Technology,
Bedford, UK*

ELSEVIER APPLIED SCIENCE
LONDON and NEW YORK

ELSEVIER APPLIED SCIENCE PUBLISHERS LTD
Crown House, Linton Road, Barking, Essex IG11 8JU, England

Sole Distributor in the USA and Canada
ELSEVIER SCIENCE PUBLISHING CO., INC.
52 Vanderbilt Avenue, New York, NY 10017, USA

WITH 7 TABLES AND 119 ILLUSTRATIONS

© ELSEVIER APPLIED SCIENCE PUBLISHERS LTD 1987

British Library Cataloguing in Publication Data

Meguid, S. A.
Integrated computer-aided design of
mechanical systems.
1. Engineering design — Data processing
I. Title
620′.00425′0285   TA174

Library of Congress Cataloging-in-Publication Data

Meguid, S. A.
Integrated computer-aided design of mechanical
systems.

Includes bibliographies and index.
1. Computer-aided design.   2. Engineering design —
Data processing.   I. Title.
TA174.M386   1986   620′.00425′0285   86-11607
ISBN-13: 978-94-010-8024-8   e-ISBN-13: 978-94-009-3409-2
DOI: 10.1007/978-94-009-3409-2

Special regulations for readers in the USA

Phototypesetting by Tech-Set, Gateshead, Tyne & Wear.

To Eli and

Jenna Danielle

Preface

In this book, the author has presented an introduction to the practical application of some of the essential technical topics related to computer-aided engineering (CAE). These topics include interactive computer graphics (ICG), computer-aided design (CAD), computer-aided analysis (CAA) and computer-integrated manufacturing (CIM). Unlike the few texts available, the present work attempts to bring all these seemingly specialised topics together and to demonstrate their integration in the design process through practical applications to real engineering problems and case studies.

This book is the result of the author's research and teaching activities for several years of postgraduate and undergraduate courses in mechanical design of rotating machinery, computer-aided engineering, practical applications of finite elements, solid mechanics, engineering dynamics and properties of materials at Cranfield Institute of Technology, Oxford Engineering Science and the University of Manchester Institute of Science and Technology (UMIST). It was soon realised that no books on the most powerful and versatile tools available to engineering designers existed. To satisfy this developing need, this book, on the use of computers to aid the design process and to integrate design, analysis and manufacture, was prepared.

The main text examines the detailed design principles of the various mechanical components which are not peculiar to the present case studies but common to other fields of mechanical engineering. The design parameters necessary for successful operation are also considered and the steps taken to ensure maximum life and reliability of the components are outlined. Although every attempt has been made to verify the different designs, no responsibility can be accepted for their performance in practice.

Following a brief introduction of the terminology used, Chapter 2 provides a detailed description of the techniques adopted in three-dimensional automated modelling. Chapter 3 provides a detailed account of the computer-aided design of two practical case studies involving the design of (i) a centrifugal peening equipment and (ii) a fluid coupling. Throughout this work, both 3-D solid modelling and 2-D draughting were interactively utilised to create, manipulate and retrieve the different geometries. In view of its importance to design, analysis and manufacture, emphasis was given to solid modelling representation. Chapter 4 provides a truly practical introduction to the application of the finite element method to structural stress analysis of mechanical systems. In Chapter 5, complete static stress analysis and dynamic response studies are performed on the different designs examined. In particular, it was desired to demonstrate the versatility of the techniques adopted in data-handling during model creation, mesh definition and model checking aspects. Chapter 6 deals with the computer-integrated manufacturing aspects of the work using the same data base developed during the modelling stage. Some coverage is also provided for numerically controlled machines.

The book is intended for mechanical, industrial and manufacturing engineers, design engineers and engineering managers. The work is also intended for postgraduates undertaking mechanical, industrial, manufacturing and production engineering degrees. It is also useful for technologists, academics and researchers working in the field of CAD/CAM. The material included is largely self-contained and readers are required to have some basic knowledge of design, analysis and manufacture; their awareness of computers will also be valuable.

Finally, I wish to acknowledge the following organisations for giving permission to reproduce a few relevant diagrams: Structural Dynamics Research Corporation — SDRC (USA), Tilghman Wheelabrator (UK) and Vacu-Blast International (UK).

S. A. MEGUID

Contents

Contents

Principal Notations

A	Area of cross-section; area of cross-section of one-dimensional element
$[B]$	Matrix containing the derivatives of shape functions
$[D]$	Elasticity matrix (relating stresses and strains)
E	Young's modulus
$E^{(e)}$	Young's modulus of the (e)th element
F	Force
$\{F\}$	Load vector
G	Modulus of rigidity
h	Convection heat transfer coefficient
I	Moment of inertia
$[J]$	Jacobian matrix
$[K]$	Global stiffness matrix
$[K^{(e)}]$	Element stiffness matrix
k	Thermal conductivity
L	Total length of bar
$\mathscr{L}$	Lagrange multiplier
$l^{(e)}$	Length of one-dimensional element (e)
N	Shape function
$\{p\}$	Load vector
r, θ, z	Cylindrical polar co-ordinates
R	Radius
T	Temperature; torque; transpose
u, v, w	Respective components of displacements in x, y and z directions
V	Volume of a body

v	Velocity
W	Work done by external forces
x, y, z	Cartesian co-ordinates; global co-ordinates
Y	Yield stress

α	Coefficient of linear thermal expansion
γ	Shear strain
$\{\delta\}$	Displacement vector
ε	Strain
$\varepsilon^{(e)}$	Strain in the (e)th element
$\{\varepsilon\}$	Strain vector
ν	Poisson's ratio
ξ, η, ζ	Intrinsic co-ordinates
π	Strain energy of a solid body
$\pi^{(e)}$	Strain energy of the (e)th element
ρ	Density
σ	Stress
$\sigma^{(e)}$	Stress in the (e)th element
$\{\sigma\}$	Stress vector
τ	Shear stress
ω	Frequency of vibration; angular velocity

Superscripts

(e)	Identifies the element number
$[\ ]^T$	Indicates the transpose of a matrix

Subscripts

s	Shot
t	Target

Chapter 1

Introduction

1.1 THE DEVELOPING DESIGN PROCESS

The modern engineer has professional obligations to clients and to society. In fulfilling these obligations, the engineer must exercise judgement, make decisions and accept responsibility for such actions. Society has become increasingly aware of and sensitive to the consequential impacts that develop due to the nature and implementation of engineering works. To satisfy their obligations to society, engineers must be able to anticipate a variety of viewpoints and requirements and to incorporate these into decision-making activities. This calls for more than mere involvement with technical and functional aspects of problems. It also requires the co-ordination of social goals and objectives and an appreciation for the continuous interactive process that occurs in planning, design and project implementation.

The engineer must perceive problems in their full environmental context and seek the solution that best satisfies the goals and objectives of clients and society. In the design process, the engineer must be aware of the environment in which the problem arises and establish relevant goals and objectives. Once the problem has been identified, it is necessary to define and model the problem as well as consider alternative solutions; provision must be made for the selection and implementation of the best acceptable solution. The degree to which an engineer is able to perform these activities within the available resources establishes his or her professional capabilities and credibility.

1

The limited traditional objective of a designer was to produce drawings for the approval of his client and for the instruction of manufacturers. In more recent years, designers are called on to seek solutions to problems that have a far-reaching impact on society. The range of possible alternative solutions must be established and evaluated in terms of improvements to the way of life, public health, safety and the availability of resources. The solution of such problems requires the careful and responsible application of scientific principles, together with a thorough understanding of the social, political and economic environment in which these problems exist.

The growing awareness of the complexity of society has compounded the engineering design process because of the breadth of issues that are involved. At the same time, society is becoming aware of and sensitive to the forces that are moulding its form and environment. Thus engineering designers need to re-evaluate their approach to problem formulation and to the manner in which their decisions, designs, manufacture and plans are presented to society.

In general, designing should not be confused with art, with science or with mathematics. It is a hybrid activity which depends, for its successful execution, upon a proper blending of all three and is most unlikely to succeed if it is exclusively identified with any one. The main point of difference is timing. Both artists and scientists operate on the physical world as it exists in the present while mathematicians operate on abstract relationships that are independent of historical time. Designers, on the other hand, are forever bound to treat as real that which exists only in an imagined figure and have to specify ways in which the foreseen object can be made to exist.

To the extent that designers need to know the present before they can predict the future, they need scientific doubt and the ability to set up and to observe the results of a controlled experiment. But when they deal with the future itself, scientific doubt is of no use and some other ingredient, nearer to religious faith, has to be employed [1.1]. The artistic approach is relevant when designers have to find their way through a vast number of alternatives while searching for a new and consistent pattern on which to base their decisions. On these occasions it is necessary to operate at the speed of thought upon a quickly responding medium. Traditionally this medium has consisted of sketches and of accurate mental pictures of tentative designs. Currently, we can expect the screen of an interactive on-line computer to provide the facility for the rapid exploration of alternative designs.

1.2 DESIGN METHODOLOGY

Design in general terms can be defined as the means by which solutions are contrived to people's problems and in response to a need. This definition draws no distinction between the traditionally separate fields of design, analysis, draughting and manufacturing. Before examining the several facets of computer-aided engineering, let us consider the general iterative method of the design process.

The realisation by someone that a need exists for solving a particular problem will lead to the broad definition of the problem in hand. This will in turn assist in the formulation of the problem in engineering terms. Effectively, the formulation of the problem would entail the compilation of a detailed design specification. This specification will generally include functional and physical characteristics, cost, quality, performance, etc. It is only at this stage that a designer employs intuition and experience to produce a preliminary solution to the problem; a certain component, sub-assembly or sub-system of the general assembly is conceptualised by the designer.

An important step in the entire process is the examination and allocation of the engineering-oriented resources that can be applied in attempting to solve the problem. These resources are associated with the amount of time in which the solution must be achieved as well as the manpower resources that can be directed towards achieving a solution. This step is critical because resource constraints may affect the approach taken to the attack on the problem area or even force the scope of the problem to be reconsidered. In both cases, the problem formulation may have to be considered.

The resulting preliminary designs are then subjected to the appropriate analysis to determine their suitability for the specified design constraints. If these designs fail to satisfy the constraints, they are then redesigned or modified on the basis of the information gained from the analysis. This iterative process continues until the proposed designs meet the specifications, or until the designer is convinced that the design is not feasible within the specified constraints. The components, sub-assemblies or sub-systems are then synthesised into the final overall system in a similar iterative manner [1.2].

The above is then followed by an assessment of the design against the specifications established during the problem definition phase. This assessment often requires the fabrication and testing of a prototype model to evaluate operating performance, quality, reliability, etc. The final phase of the design activity is the documentation of the design by means of engineering drawing, material specification, list of components, etc.

Engineering design has traditionally been accomplished on drawing boards, with the design being documented in the form of detailed engineering drawings. Effectively, mechanical design includes the drawing of the complete product as well as its components and sub-assemblies, tools and fixtures necessary for the manufacture of the product.

Clearly, the design process relies entirely on the iterative approach, in the sense that each iteration provides an improvement in the product. The main problem with this iterative approach is that it is time consuming and many hours of attempts are required before a design project is completed.

Figure 1.1 shows a schematic of the iterative design method.

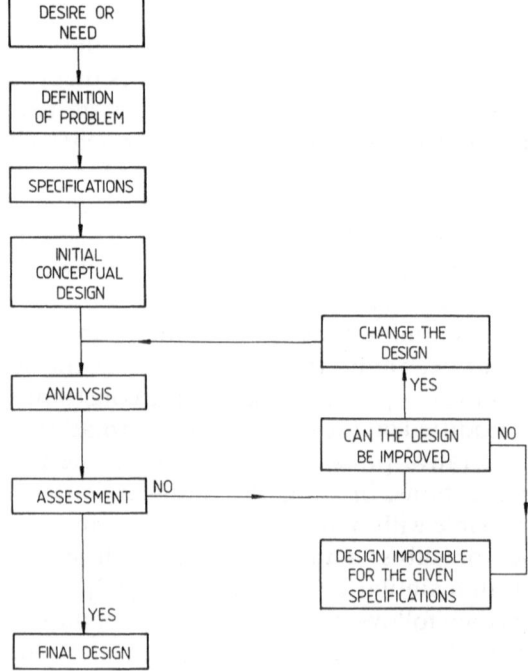

FIG. 1.1. Iterative design methodology.

Today, a number of design-related tasks can be performed with the assistance of a computer. These tasks include

(i) interactive computer graphics and modelling,
(ii) computer-aided analysis,
(iii) computer-aided draughting,
(iv) computer-integrated manufacturing.

They are explained in detail in the following sections.

The benefits of employing the modern computer to aid the design and manufacture processes can be enormous. They include improved engineering productivity, shorter lead time, reduced labour, improved accuracy of designs, savings in materials and machining time, faster response to changes in designs and/or assemblies, integration between designers, analysts and production engineers and improved documentation

1.3 INTERACTIVE COMPUTER GRAPHICS (ICG)

Images in one form or another generated by computer are in evidence in our daily existence; they are performing a greater diversity of functions and are about to change the way we live. Indeed, computers are very powerful tools and vision is a very powerful medium; the union of the two is a revolutionary means of communication and creativity and is known as computer graphics [1.3].

The nature and quality of an image is obviously dependent upon the machine which creates it. The three categories of machines used in interactive graphics are

(i) main-frame computer,
(ii) mini-computer,
(iii) micro-computer.

The main-frame computer is distinguished by its cost, capacity and function. The price of a new general purpose main-frame computer can run into millions of pounds. The main memory capacity is much larger than the mini-computer, and the speed with which computations, including calculating the exact intensity and colour of each one of millions of points of light and assembling them to form objects of stunning detail and realism, can be made is much faster than mini- and micro-computers.

At the other end of the scale, a micro-computer can be used. Despite the small size of its circuitry, the micro-computer is capable of performing all the regular functions of a traditional computer. Because of its small size, the cost of a micro-computer is much lower than the cost of either a mini-computer or a main-frame computer.

Mini-computers often overlap with both micro-computers and main-frame computers. In fact, mini-computers can be utilised for the same general functions as a large main-frame computer. However, the size of the job must be smaller to be within the storage capacity of the machine.

Although computer graphics is a young discipline, it is growing so rapidly that it is beyond the scope of the present work to explore its full potential. Consequently, only the relevant examples are presented in the following sections for their imaginative content, techniques adopted in creating the object and for their practical and visual appeal.

Interactive computer graphics (ICG) comprises the following important functions [1.4]:

  (i)   modelling, which is concerned with the description of an object in terms of its spatial co-ordinates, lines, areas, edges, surfaces and volume,
 (ii)   storage, which is concerned with the storage of the model in the memory of the computer,
(iii)   manipulation, which is used in the construction of the model from basic primitives in combination with boolean operations,
 (iv)   viewing, in this case the computer is used to look at the model from a specific angle and presents on its screen what it sees.

The above functions hold true throughout the computer graphics process, but there are many areas of specialisation within the subject area. For instance, the importance of a fast and effective *interactive* dialogue between the human user and the machine is stressed as in video games. On the other hand, in the quest for reliable, economical and creative designs the use of interactive computer graphics cannot be over-emphasised. The enormous economic advantages which accrue from the utilisation of computers in the design of engineering components, assemblies and plants have been a major driving force behind the development of ICG. One rapidly developing area is man–machine interaction. An important question in industrial design is how can the designer make best use of his time? The answer to this question involves not just the use of computers, but more effectively how well a human being can integrate with them.

1.4 COMPUTER-AIDED ENGINEERING (CAE)

Computer-aided engineering is a term embracing the related areas of computer-aided design (CAD), computer-aided analysis (CAA) and computer-integrated manufacturing (CIM). They are in addition to the many supporting activities, such as planning, management, and control of the manufacturing plant through either direct or indirect computer interface. CAE is a combination of techniques in which man and machine are blended into a problem solving team, intimately coupling the best characteristics of each. The result of this combination works better than either man or machine would work alone, and by using a multi-discipline approach it offers the advantages of integrated team work [1.5].

1.4.1 Computer-Aided Design (CAD)

Computer-aided design involves any type of design activity which makes use of the computer to create, develop and modify an engineering design. In CAD the model which is created by ICG and stored in the computer memory will provide the data base necessary for subsequent operations such as the analysis and the manufacture.

CAD implies by definition that the computer is not used when the designer is most effective and vice versa. This being so, it is therefore useful to identify which processes can best be separately performed by each, and where one can aid the other. It can be seen that in most cases the two are complementary, that for some tasks man is far superior to the computer, and that in others the computer excels. It is therefore the marriage of the characteristics of each which is so important in CAD. These characteristics affect the design of a CAD system in the following areas:

   (i)   Design construction logic — the method of constructing the design.

  (ii) Information handling — the storing and communication of design information.

 (iii) Modification — the handling of errors and design changes.

 (iv) Analysis — the examination of the design and factors influencing it.

Generally, computer modelling can be performed using any of the following approaches: (i) wire-frame, (ii) bounding surfaces, and (iii) full solid definition. A summary of the advantages and disadvantages of these three different approaches is provided in Chapter 2.

There are several fundamental reasons for implementing a computer-aided design system. They include

(i)　increased productivity of the designer through the visualisation of the product and its component, sub-assemblies and related parts and by reducing the time necessary for the development of a conceptual design, analysis and documentation,

(ii)　integration between design, analysis and manufacture through the provision of a common data base,

(iii)　design error can be eliminated at the early stages of design,

(iv)　improved documentation and standardisation of engineering drawings,

(v)　reduction and/or elimination of mundane and repetitive jobs.

1.4.2 Computer-Aided Analysis (CAA)

Much product development down through the years has been evolutionary, in the sense that many of today's designs are the result of a great deal of trial and error experience. One design idea performs better than another and is selected for incorporation in future versions of a product. Similarly, scaling of a product's physical parameters is used when a proven design is applied to an increased loading environment. This experience and the intuitive engineering approach has served us well and will continue in many, if not most, areas of mechanical design.

The modern engineer/designer while making use of these traditional methods will increasingly need to involve himself in utilising sophisticated analysis techniques to enable him to produce not only creative, imaginative and economical designs, but also reliable, safe and robust designs capable of withstanding the applied mechanical and thermal loads.

Computer-aided analysis involves any type of analysis activity which makes use of the computer to aid the design process. The analysis may involve structural integrity (stresses and strains) calculations, dynamic response calculations and/or heat-transfer computations.

Undoubtedly, the most powerful analysis feature of any computer-aided system is the finite element method (FEM) [1.6]. With this technique, complete structural stress analysis, dynamic response calculations, and heat transfer of a given component and/or design can be accurately assessed. In view of its importance to the design process, a complete chapter is devoted to the basic foundation of the finite element method, relevant applications and some of the imposed limitations.

Figure 1.2 shows a three-dimensional finite element model of a component used in the design of a centrifugal peening equipment ready for the stress and strain analysis using the appropriate analysis package. It must be pointed out that if the finite element analysis indicates some weakness in the design, the designer can modify the shape of the component and re-analyse the modified design.

In addition to the above, the analysis of mass properties is another feature of any CAE system. It provides properties of a solid object being analysed, such as surface area, weight, volume, centre of gravity, area and mass moments of inertia about the appropriate axes.

FIG. 1.2. A three-dimensional finite element model using interactive mesh generation.

1.4.3 Computer-Aided Draughting (CAD)
This involves the generation of hard-copy engineering drawings directly from the CAE data base. In some early CAD departments, the automation of the draughting process was the main justification for investing in a CAD system. In fact, the utilisation of CAD systems can

increase the draughting productivity by approximately four times over conventional manual draughting [1.7].

Some of the graphics features of CAD systems lend themselves well to the draughting process. These include automatic dimensioning, cross-hatching, scaling of drawings, development of sectional views and enlarged view of a particular component. The ability to rotate, translate, cut and zoom can be of significant assistance to the designer.

Figure 1.3 shows an engineering drawing with four views displayed which were produced by a CAD system. Note how much the isometric view provides a higher degree of understanding of the object especially for those who cannot think in three dimensions.

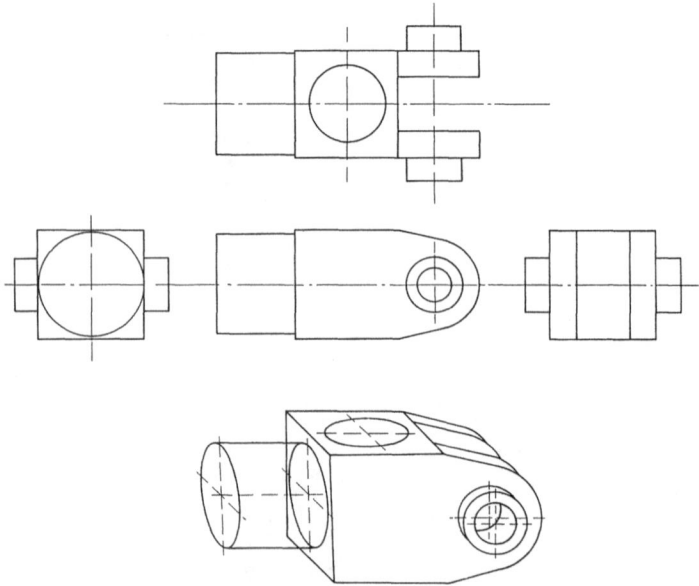

FIG. 1.3. Engineering drawing with four views generated automatically by a CAD system.

1.4.4 Computer-Integrated Manufacturing (CIM)

In an integrated CAE system, a direct link is established between product design and manufacturing. Indeed, it is the goal of CAE not only to automate the different phases of the traditional design process, but also to integrate design and manufacture. This will undoubtedly reduce the time consumed in the duplication of effort by the designers and production/manufacturing engineers.

The manufacturing data base is an integrated CAE data base. It includes all necessary data on the product generated during the conceptual design process, which includes geometry data, materials used, parts lists, etc. In addition, further information is provided for the manufacturing of the components. The manufacturing data base can then be used in conjunction with computer numerically controlled (CNC) machines for the accurate manufacturing of very complex geometries and for mass production [1.8].

1.5 INITIAL GRAPHICS EXCHANGE SPECIFICATION (IGES)

Fundamental incompatibilities amongst entity representation greatly complicate data base interchange between CAD/CAM systems. Even simple geometric entities such as circular arcs are represented by incompatible forms in many of the current systems. Now that computer data bases are used in defining product geometry for all phases of design, analysis and manufacture, effective procedures for interchanging these data bases are becoming increasingly important.

The data base interchange problem is complicated by the complexity of CAD/CAM systems, the varying requirements of organisations using them, the restrictions on access to proprietary data base information and the rapid pace of technological advancement in the field. The initial graphics exchange specification, which is part of the new ANSI Y14·26M standard [1.9] and [1.10], now provides a solid foundation for data base interchange in the form of a standardised file organisation and standardised representations for CAD/CAM data base entities.

The IGES specification currently includes representations for some entity types found in any CAD/CAM systems. These entities include geometrical features such as points, lines, arcs and ruled surfaces; annotation features such as dimensions and notes; and structural features such as drawings, macros and properties and views. IGES reduces the data base interchange problem to the design of (i) a pre-processor to map the original data base into an IGES file and (ii) a post-processor to map the IGES file into the desired data base. Figure 1.4 shows the data base transmission using IGES.

It is unfortunate that the situation with regard to CAD data interchange is currently very confusing. Three versions of IGES exist. 'SET' is a standard in France and 'VDAFS' in Germany. The Americans are already developing Product Data Exchange Standard 'PDES', intended as a successor to IGES, and five European countries, including Britain, are currently working together under the EEC–ESPRIT Scheme to develop a co-ordinated European approach. A unified international standard should result from these efforts in the near future.

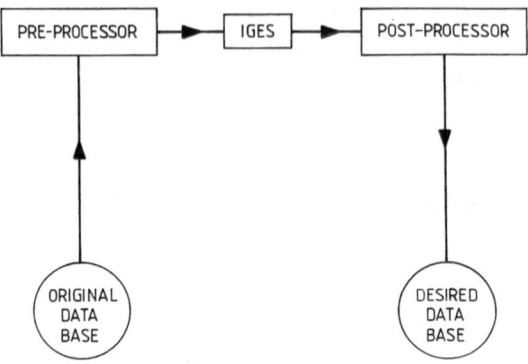

FIG. 1.4. Data base transmission using IGES.

REFERENCES

[1.1]   J. C. Jones, *Design Methods: Seeds of Human Futures*, Wiley International, London, 1970.

[1.2]   J. E. Shigley, *Mechanical Engineering Design*, 3rd Edition, McGraw Hill, New York, 1977.

[1.3]   S. H. Chasen, Historical highlight of interactive computer graphics, *Mechanical Engineering*, pp. 32–41, November, 1981.

[1.4]   J. D. Foley and A. Van Dean, *Fundamentals of Iterative Computer Graphics*, Addison-Wesley, Massachusetts, 1983.

[1.5]   J. K. Krouse, CAD/CAM — Bridging the gap from design to production, *Machine Design*, pp. 117–25, June, 1980.

[1.6]   O. C. Zienkiewicz, *The Finite Element Method in Engineering Science*, McGraw Hill, London, 1977.

[1.7]   J. C. Lange and D. P. Shanahan, *Interactive Computer Graphics Applied to Mechanical Drafting and Design*, John Wiley & Son, New York, 1984.

[1.8]   M. P. Groover and E. W. Zimmers, Jr, *CAD/CAM Computer-Aided Design and Manufacturing*, Prentice Hall Inc., Englewood Cliffs, NJ, 1984.

[1.9]   R. W. Nagal, W. W. Braithwaite and P. R. Kennicott, Initial Graphics Exchange Specification, version 1.0, National Bureau of Standards Publication NBSIR, 30, 1978.

[1.10] J. W. Lewis, Specifying and verifying IGES processors, *Proc. Conference on CAD/CAM Technology in Mechanical Engineering*, 24–26 March, Massachusetts Institute of Technology, Cambridge, Mass. pp. 377–98, 1982.

Chapter 2

Three-Dimensional Automated Modelling

2.1 AUTOMATED COMPUTER GRAPHICS

The engineering drawing is a key document in any design/manufacturing activity. The information it contains is the input to the majority of other related functions which range from metal cutting and forming to the purchase of materials and components. As the drawing stores formed part of the organisational core data in the pre-computer era, drawings or more importantly object description/product definition form the core to the engineering data base. Though the 'paperless' organisation may soon become a reality, it is doubtful if a 'pictureless' organisation will ever evolve in the manufacturing industries.

Although engineering draughting has served us well in the past, current efforts towards the integration between computer-aided design and computer-aided manufacturing have shown that it has serious limitations as a means of geometric object description or definition for the future. The problem is essentially that drawings are understandable by humans but not by computers. Integrated CAD/CAM develops and utilises a common data base for both the design and manufacture stages of the object.

Automated computer graphics is a way of converting the computer's impulses into engineering documents and, conversely, to translate the designer's instructions into electronic data [2.1]. In many of the more sophisticated systems, we need to know little about computer programming in order to interactively develop and control the human–machine effort. In general, automated computer graphics include any device which converts computer language to a common human language, and vice versa, with the intent of solving design problems by creating graphical images. An example of an automated system and how it works can be studied by examining the block diagram of Fig. 2.1.

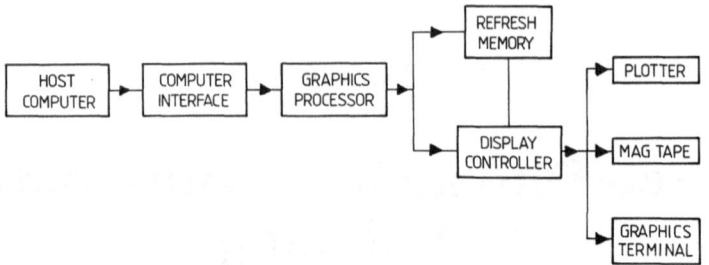

FIG. 2.1. Typical automated graphics system.

2.2 DIFFERENT TYPES OF MODELLING

Essentially, the integration between CAD and CAM necessitates the use of some means of object definition other than the drawing. One such means relies on the construction of a common data base using known mathematical terms in the representation of the object. More recent CAD systems possess the capability to define objects in three dimensions. This allows the designer to develop a full three-dimensional model of an object in the computer rather than a two-dimensional illustration. The computer can then generate orthogonal views, perspective drawings and close-up details of the object. More recent developments in this area include three-dimensional wire-frame, surface and solid modelling techniques [2.2]. These three methods of object definition are described below.

2.2.1 Wire-frame modelling

A common form of 3-D modelling capability found in many CAD systems is wire-frame. As its name implies, a wire-frame object has no surfaces attached to it. This enables the definition of only the edges of an object. In this case the object is transparent; thus anything but the simplest model is very difficult to interpret visually.

Two types of wire-frame modelling exist: (i) 2½-D and (ii) 3-D modelling. In the 2½-D modelling approach, the two-dimensional drawing which is generated in CAD is stored within the computer as an assembly of lines and curves. The geometrical details of each entity are recorded, together with some information about their connectivity. If we consider the plane of the drawing as the *xy*-plane, it is then possible to obtain a simple representation of certain three-dimensional objects simply by associating depth information with each line or curve, in the form of a *z*-value. Figure 2.2 shows an example in which the end points of all the lines and curves in the original drawing become vertical edges in three dimensions. In this case the nature of the surfaces of the associated solid object are clear; a line in the drawing becomes plane, a circular arc becomes a part of a cylinder, etc., either in the vertical or horizontal planes.

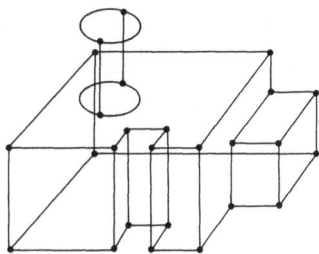

FIG. 2.2. A 2½-D wire-frame representation of a composite block with a cylindrical boss.

Most of the turnkey CAD systems have gone further to allow for the 3-D wire-frame definition of objects. Whereas in 2½-D wire-frame modelling all edges are either horizontal or vertical, this is no longer the case for 3-D wire-frame modelling. Unfortunately, one of the most important features of the 2½-D modelling approach is lost; the system can no longer accurately detect the nature of all the surfaces of the corresponding solid object from the edge data which it stores.

The basic 3-D wire-frame object definition suffers from a number of disadvantages. Section property and mass calculations are impossible, since the object has no faces attached to it. Graphical aids to visualisation are not possible. In fact the wire-frame object does not necessarily correspond to a unique solid object. Figure 2.3 shows a wire-frame object with two corresponding views of solid objects.

A partial solution to the problems listed above and favoured by some CAD graphics systems is the introduction of implicit or explicit attachment of surfaces to the wire-frame objects. This improves visualisation of the model through 'hidden line removal'. As the name implies, this automatically deletes all those lines that would not be seen when viewing the model from a given direction. It must be realised, however, that the resultant model of the object is view-dependent and without the appropriate surfaces and volumes. It is therefore well short of satisfying the requirements of true engineering design.

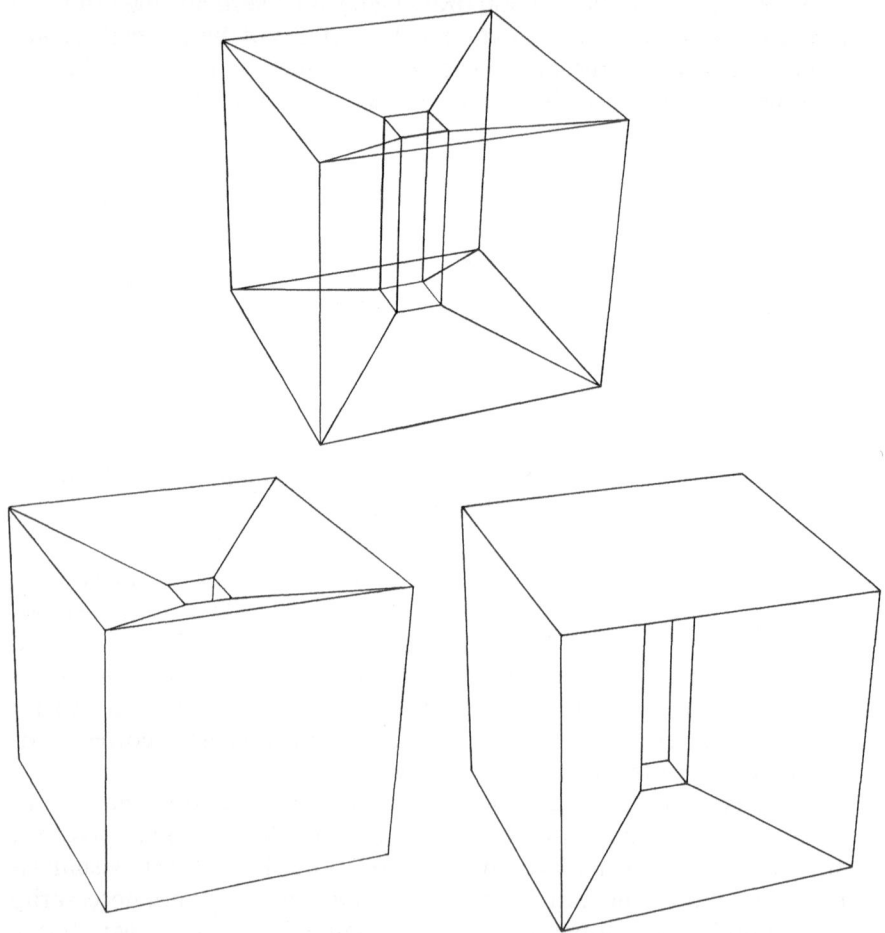

FIG. 2.3. A wire-frame object with two corresponding solid objects.

2.2.2 Surface modelling

Surface modelling originated from techniques adopted by the aerospace industry some 45 years ago where conic sections were used to define the appropriate curves. These curves were then analytically blended to give a smooth surface. In recent years, some CAD systems have adopted the lofting approach in which the designer works initially in terms of plane cross-sections of the desired surface (Fig. 2.4(a)) which are then blended by the computer to give a smooth surface containing all the cross-sectional curves as shown in Fig. 2.4(b) [2.3].

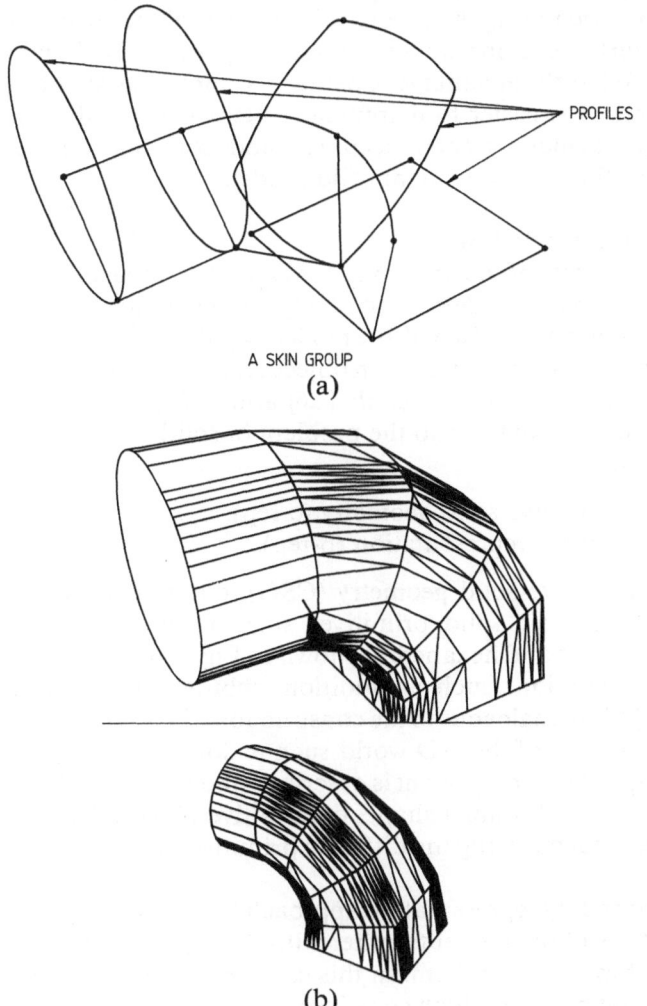

(a)

(b)

FIG. 2.4. Surface modelling using cross-sectional curves (After Ref. [2.3]).

Other CAD systems have generalised the lofting approach. These allow the construction of a surface which is defined with respect to a set of specified points in space. In this case, the definition is provided by the bounding surfaces of a 3-D object; further details of the different techniques adopted in surface modelling can be found in Refs [2.4] and [2.5].

Visually the model is solid and will facilitate hidden line removal and in some instances shaded images. In reality, however, it still lacks substance and sectioning the model will only reveal the edge definition of the cut surfaces. Consequently, mass properties calculations are not possible. It has the advantage of being less demanding computationally and therefore is easier to manipulate interactively and may, in some situations, achieve greater accuracy and sophistication of surface definition than an equivalent solid model.

2.2.3 Solid modelling

The current state of the art in 3-D CAD is 'solid modelling' which provides a full solid 3-D definition of objects, generating in effect a computerised equivalent of a real-life model. A solid model will involve both surface and edge definition of an object and will furthermore embody a recognition of volumetric details [2.6] and [2.7].

Two basic approaches to the problem of solid modelling have been developed:

(i)   constructive solid geometry (c-rep),
(ii) boundary representation (b-rep).

In constructive solid geometry (CSG), the designer constructs his model from basic solid primitives such as blocks, cones, spheres, cylinders, tubes and hexahedra (shown in Fig. 2.5) in combination with boolean operations such as addition, subtraction, union and intersection. This is analogous to the construction of a 2-D drawing using the basic primitives of the 2-D world such as lines and arcs. In CSG, the final shape of the component is described and maintained internally by a tree structure of simpler shapes of primitives. Figure 2.6 shows the different steps taken, using the different primitives, in constructing a solid model.

The boundary representation approach keeps a list of the faces, edges and vertices of the model together with the topological and adjacency relationships between them. In this case, the topology is used to determine the set of edges which constitute the boundary of a particular face or which meet at a specific vertex. In the b-rep, the designer can construct his model from drawing (i) the outline or the profile of the object or (ii) the various views of the object containing the interconnecting lines among the views to establish their connectivity. These techniques are depicted in Figs 2.7 and 2.8. Figure 2.7 shows the two-dimensional profile of an axisymmetric object which was used to

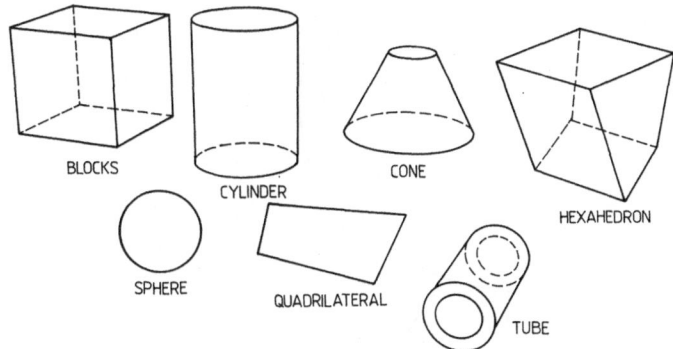

FIG. 2.5. Object modelling library of primitives (After Ref. [2.3]).

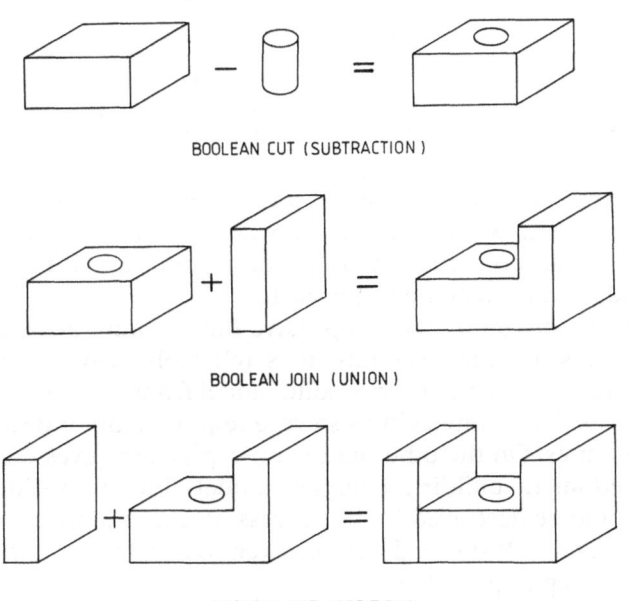

FIG. 2.6. Effect of some boolean operations on primitives.

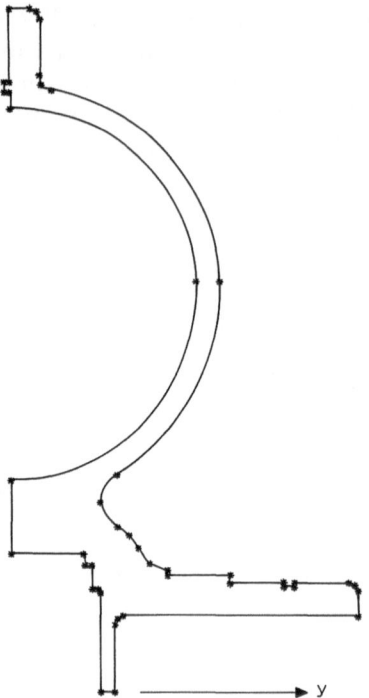

FIG. 2.7. A 2-D 'axisymmetric' profile used in b-rep modelling of an object.

construct the corresponding solid model from (usually) 360° rotation
about the y-axis, while Fig. 2.8 provides the different input views
necessary for the construction of the corresponding solid model using
the boundary representation approach.

Both methods (b-rep and c-rep) have their relative advantages and
disadvantages. In c-rep systems, it is relatively easy to construct a
topologically correct and precise solid model from the available library
of primitives. It is compact in its storage requirements, but slow at producing pictures. On the other hand, a b-rep system gives the designer
more freedom in building complex models but the validity of the
models could be destroyed in the process. It is also more expensive on
memory. Figure 2.9 shows the steps taken to construct an object using
b-rep and c-rep approaches.

With some b-rep systems, some 'tweaking' or local modification is
possible — so long as the topology of the model is unaffected. This

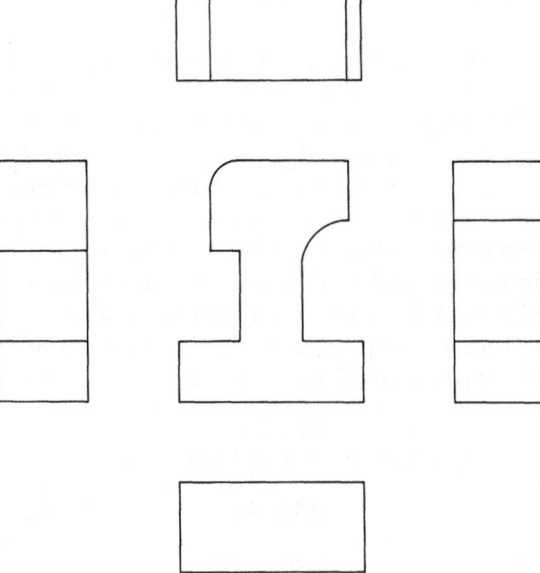

FIG. 2.8. Input views of the types required for b-rep modelling of an object.

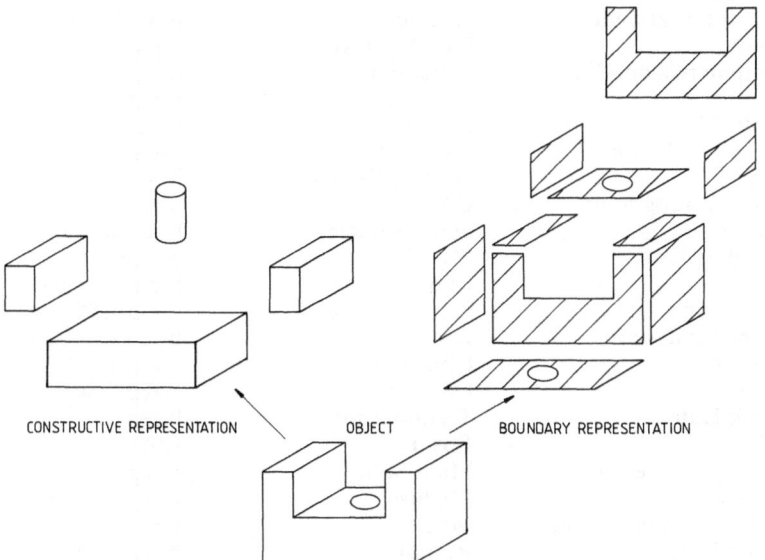

CONSTRUCTIVE REPRESENTATION   OBJECT   BOUNDARY REPRESENTATION

FIG. 2.9. Comparison between b-rep and c-rep modelling techniques.

technique can be used to chamfer or fillet a sharp edge or move a hole within a face.

Most commercial systems use a hybrid approach, converting the CSG working model to a b-rep definition, and store both models in memory. The CSG might then be used for mass property calculations and the b-rep for down loading edge data for producing 2-D drawings.

Most solid modellers on the market today (approximately 30 commercially available solid modellers) derive from two research projects; the PADL project at the University of Rochester and the BUILD project at the University of Cambridge. Table 2.1 provides examples of the commonly used solid modellers and their representation.

Some b-rep programs approximate curved surfaces by using facets. The size of the facet is traded-off against speed of interaction or accuracy

Table 2.1
Commonly Used Solid Modellers

Vendor	Modeller	Representation method
Applicon	Solids Modeling II	CSG/faceted
Auto-trol	BMOD, CMOD	b-rep/CSG
CADCentre	DIAD	CSG
CAE International	Geomod	b-rep/faceted
Calma	DDM/Solids	b-rep
Computervision/CIS	Solidesign	b-rep
	Medusa	b-rep/faceted
Control Data	ICEM	CSG
Data Translation	Polycad/10	b-rep
ECS	Graftek Comet	b-rep
Ferranti Infographics	CAM-X	b-rep
Gerber	Solid Modeler	b-rep/faceted
Gould SEL	MCAD	b-rep
IBM	Catia	b-rep/faceted
Intergraph	Solid Modeler	b-rep/CSG
Matra Datavision	Euclid	b-rep/CSG/faceted
McAuto	Unisolids	b-rep/CSG
MCS	Omnisolid	b-rep/CSG
Norsk Data	Technovision	b-rep
Pafec	Boxer	CSG
Perspective Design	MicroSolid	b-rep
Prime	Medusa	b-rep/faceted
Radan Computational	Vole	CSG
Shape Data	Romulus	b-rep
Sperry	Solid Modeler	b-rep/CSG
Techex	Cubicomp	b-rep

required. Preliminary designs can use large facets for speed and these can be reduced right down when the design has been finalised, for increased accuracy.

Faceted models are not suitable for all applications, since however small the facets, the model is still inexact. Some modellers store a precise model in the computer but use a faceted representation for display purposes. For really fast interaction, designers work in wire-frame producing hidden-line or colour-shaded images pictures only when necessary.

2.3 MERITS OF SOLID MODELLING

Solid modelling provides the benefits of designing in 3-D, a far more natural mode of expression once mastered. It can be used at many different stages in the design and manufacture of a product. At the conceptual design stage, it can provide a visual aid, possibly replacing the prototype.

Three-dimensional solid modelling allows sectional views at any position or angle, and exploded views are also possible. The 3-D geometry can be transferred for structural stress analysis, dynamic response studies and for the generation of numerically controlled machining data.

Solid modelling also allows for interference checks; design errors which could be catastrophic at the machining or assembly stage of manufacture can be pre-empted by simulation of the assembly. The colour-shaded images pictures produced during modelling can be used for sale, marketing and technical illustrations.

Solid modelling produces a text file which contains the mass property values for the object. The specific geometry related quantities calculated by most systems include volume, weight, centre of gravity, mass moments of inertia, axial radii of gyration, products of inertia, polar radius of gyration and direction cosines of the principal axes and the axial moments of inertia.

2.4 LIMITATIONS OF SOLID MODELLING

In view of the complexities of the geometries generated using three-dimensional modelling, it usually takes longer to construct an object or model in three dimensions than to produce the equivalent 2-D views necessary for a conventional manufacturing drawing. Furthermore, the computational demands of solid modelling can significantly reduce the response performance of the system and consequently affect other users.

The use of a colour-shaded image is an important facet of three-dimensional modelling. However, if one requested view does not quite show the details required, then another 2 or 3 hours of computer processing time may be needed before seeking the next view.

Frequently, when problems occur during processing of a model, the program will fail without any indications of the reasons.

REFERENCES

[2.1]   W. K. Giloi, *Interactive Computer Graphics,* Prentice Hall Inc., Englewood Cliffs, NJ, 1978.

[2.2]   M. D. Schussel, The mechanical design process using solid modelling, *Proc. Conference on CAD/CAM Technology in Mechanical Engineering,* 24–26 March, Massachusetts Institute of Technology, Cambridge, Mass., pp. 139–45, 1982.

[2.3]   *GEOMOD User Manual,* Structural Dynamics Research Corporation (SDRC), General Electric, CAE International, USA, 1983.

[2.4]   M. J. Pratt, 3D geometric product definition for CAD/CAM: the present state and future trends, *Proc. Conference on Effective CAD/CAM,* 29–30 June, Pembroke College, Cambridge, UK, pp. 71–8, 1983.

[2.5]   I. D. Faux and M. J. Pratt, *Computational Geometry for Design and Manufacture,* Ellis Horwood Ltd, Chichester, UK, 1979.

[2.6]   W. Myers, An industrial perspective on solid modelling, *IEEE Computer Graphics,* pp. 86–97, March, 1982.

[2.7]   J. W. Boyse and J. E. Gilchrist, GM Solid: Interactive modelling for design and analysis of solids, *IEEE Computer Graphics,* pp. 27–40, March, 1982.

Chapter 3

Computer-Aided Design

3.1 INTRODUCTION

Engineering design was one of the first contenders for the sensible application of automated computer graphics (ACG). However, early work in this area indicated that it was much harder than anticipated to obtain useful results to support or speed the design process. The problem was how to construct a unified data base to hold the details of a complex design such as the fuselage of an aircraft or the hull of a ship.

The term computer-aided design means a number of things to the industrial community. In this text, computer-aided design is defined as the use of computers to aid in the construction, modification and evaluation of a design. The computer systems consist of the hardware and software. The CAD hardware would typically include the processor, graphics display terminals, keyboards and other ancillary equipment. The CAD software would typically consist of computer programs for modelling, draughting, manipulating and creating the appropriate data base for the different designs.

It must be remembered, however, that interactive computer graphics is only a tool used by the designer to speed calculations, improve the visual display capabilities and enable the storage of large amounts of data. The process of independent thinking, conceptualisation and creativity is performed by the designer.

This chapter is intended to provide an insight into the different technical aspects of computer-aided design using 2-D draughting and 3-D solid modelling. These aspects, which are closely associated with conceptual design, include automated computer modelling, synthesis and design evaluation. It is believed that these aspects can best be demonstrated by application to real engineering problems. Accordingly, two major case studies have been considered throughout this work. The first deals with the detailed design of a centrifugal shot-peening equipment, while the second deals with the detailed design of a hydraulic or fluid coupling. The main text examines the detailed design principles of the various mechanical components which are not peculiar to the present case studies but rather common to other fields of mechanical engineering. The design parameters necessary for successful operation are also considered and the steps taken to ensure maximum life and reliability of the components are outlined.

First Case Study: Computer-Aided Design of Centrifugal Shot-Peening Equipment

3.2 THE SHOT-PEENING PROCESS

Shot-peening is a cold-working process used mainly to improve the fatigue life and corrosion resistance of metallic components. The result is accomplished by bombarding the surface of the component with small spherical shots of hardened cast-steel, conditioned cut-wire or glass-beads at a relatively high velocity. After contact between the shot and a target has ceased, a small plastic indentation will have been made causing stretching of the top layers of the exposed surface. The outcome of this treatment is two-fold:

(i) the cold-working of the exposed surface layers,
(ii) the introduction of compressive residual stresses at and near the treated surface.

Both of these effects are highly effective in preventing premature failure under conditions of cyclic loadings, since fatigue failure generally propagates from the free surface of a component and starts in a zone which is subjected to tensile stresses; for further details regarding the process see Refs [3.1] to [3.4].

The primary justification for shot-peening is that its use would allow engineering components to be employed at relatively high stress levels under cyclic loading and/or aggressive environments. In the case of the aerospace industry, this means a reduction in structural weight for a specified reliability level. The treatment has proved very useful in combatting fretting, stress corrosion cracking and corrosion fatigue problems. Typical aerospace components which are currently being treated include the highly stressed regions of compressor and turbine discs and blades, many of the components of main and nose landing gear assemblies, propellor and harmonic drives and main rotor spindles. The technique has also been used to shot-peen-form wing skins to specified curvatures without resorting to conventional metal-forming techniques.

In automotive applications, it means relatively small, low cost components can be upgraded for conservative operation at stress levels that would represent poor practice without shot-peening. Springs, gears of all types and shapes, connecting rods, crank shafts and torsion bars are examples of components that can be upgraded without the use of costly alloys or increased sections.

The ability to upgrade the mechanical properties of a component by peening offers obvious opportunities in the correction of under-sized components, when fatigue failures occur after a product is standardised or in field service.

One of the techniques currently employed in treating relatively large areas and ensuring complete and uniform coverage utilises centrifugal peening equipment in which a rotating bladed wheel imparts a part of its rotational energy into providing the media with the linear momentum necessary for the impingement of the treated components.

Within the impact treatment industry, it is recognised that the wheel-assembly is the heart of any centrifugal peening installation and as such is the prime factor which determines the operating costs and efficiency. The wheel-assembly also represents the main problem area and therefore justifies closer examination.

There exist a number of interesting areas where it was considered that some computer-aided design investigations could be attempted in order to provide a more efficient design of the wheel-assembly, thus resulting in (i) a lower overall cost, (ii) an improved structural integrity of the critical components, (iii) an enhanced overall dynamic performance of the assembly and (iv) a design which can be simply and cheaply assembled and easily maintained.

3.3 HISTORICAL BACKGROUND

The first patent for '... , cutting, grinding, etc ...' by blasting with steam, water and compressed air was filed by Tilghman [3.5] in 1870. This patent included methods of using direct pressure, suction (syphon system) and partial vacuum. In the same year, Tilghman filed a patent [3.6] covering projection of abrasives by means of centrifugal force. He visualised the use of a wheel in which the abrasive was thrown by 'slider' action and by a 'batter' technique as shown in Fig. 3.1 (Ref. [3.3]).

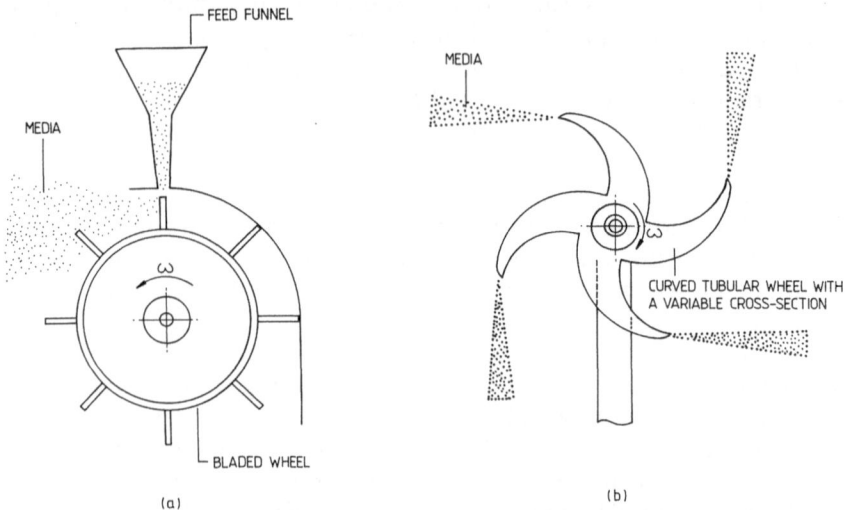

FIG. 3.1. The first blast wheels patented by Tilghman (1870): (a) 'batter' technique of media propulsion; (b) 'slider' technique of media propulsion.

The growth of the process over the early years was fitful; it was mainly looked upon as merely a useful cleaning method. However, in 1927 Herbert [3.7] devised equipment that was described as projecting a 'cloudburst' of steel balls against a metal surface for the purpose of hardening the exposed regions. No mention was made by Herbert of the possible improvement to the fatigue resistance of the treated components. Indeed, it was Weibel in 1935 [3.8] who recognised the value of the shot-peening treatment in connection with fatigue resistance of highly stressed components. There has been, ever since, considerable research and investigations into the field of shot-peening and its effect on fatigue, corrosion fatigue and stress corrosion cracking; see, for instance, Refs [3.9] to [3.12] and the references listed in them.

In these publications very little emphasis, if any, was devoted to the design of centrifugal peening equipment and its effect upon the performance of the treatment. It was therefore thought desirable to undertake the present design study.

Figure 3.2 illustrates the essential features of a typical centrifugal peening equipment. In this type of installation it is normal to feed the media by gravity through a feed spout to a distributor known as the impeller. The media stream is then picked by the rotating blades and is accelerated to the required final exit impinging velocity.

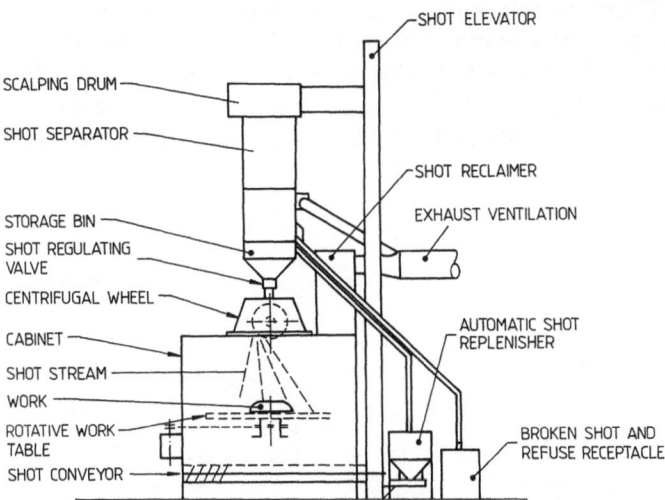

FIG. 3.2. A schematic of the general assembly of a gravity-fed centrifugal peening equipment.

3.4 DESIGN CONSTRAINTS AND REQUIREMENTS

The following design constraints were imposed on the current design of the wheel-assembly:

    (i)   range of yield strength of target materials : 200–1000 MPa
   (ii) minimum surface area to be peened   : 65 mm × 800 mm
  (iii) permissible input power   : 25 kW
  (iv) mass flow rate   : 4 kg/s

Obviously, the above mentioned design constraints are in addition to the conventional ones of being safe, easy to maintain, reliable and economical to manufacture.

3.5 CONCEPTUAL DESIGNS

A fundamental design study based on reliable and advanced analytical techniques, as well as our experience with the currently available equipment, was undertaken; one of our objectives was to improve upon the existing designs. In order to minimise the project risks and ensure mechanical integrity, detailed strength as well as dynamic response calculations were performed on the critical components of the present design. These calculations are discussed in Chapter 5.

A first attempt was made using a single-disc configuration in which eight dove-tailed straight blades were secured in position using clamping screws as illustrated in Fig. 3.3. In order to commence the design process, an initial estimate of the dimensions of the single disc together with the attached blades was made.

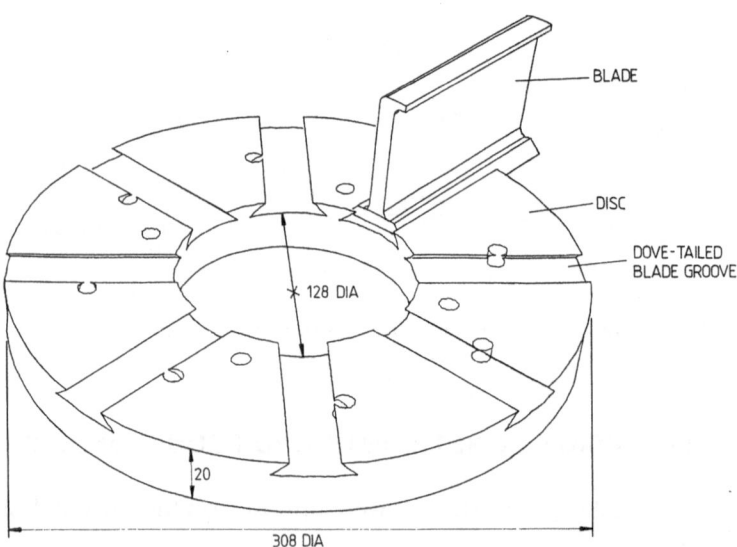

FIG. 3.3. Three-dimensional solid model representation of an eight-bladed single-disc configuration showing a blade secured in appropriate position, dove-tailed grooves and clamping holes with all hidden lines removed (dimensions in mm).

The individual components of the single-disc arrangement were developed using the hybrid approach as follows: the disc was created from a tubular primitive with the appropriate dimensions as shown in

Fig. 3.4(a). The eight dove-tailed grooves in the disc were formed by eight boolean cut operations of the hexahedron primitive of Fig. 3.4(b). The positioning holes in the disc were also performed by a boolean cut operation of a cylindrical primitive. The 2-D profile of Fig. 3.4(c) was extruded to generate a solid model of the blade. For reasons of clarity, only one blade is attached to the disc as shown in Fig. 3.3.

The above design suffers from the following shortcomings: (i) lack of rigidity of blade fixing, (ii) increased turbulence to the jet stream, which has the effect of carrying an excess of shot into the wheel-hood casing, thus causing undue wear of the protective lining, (iii) inefficient use of

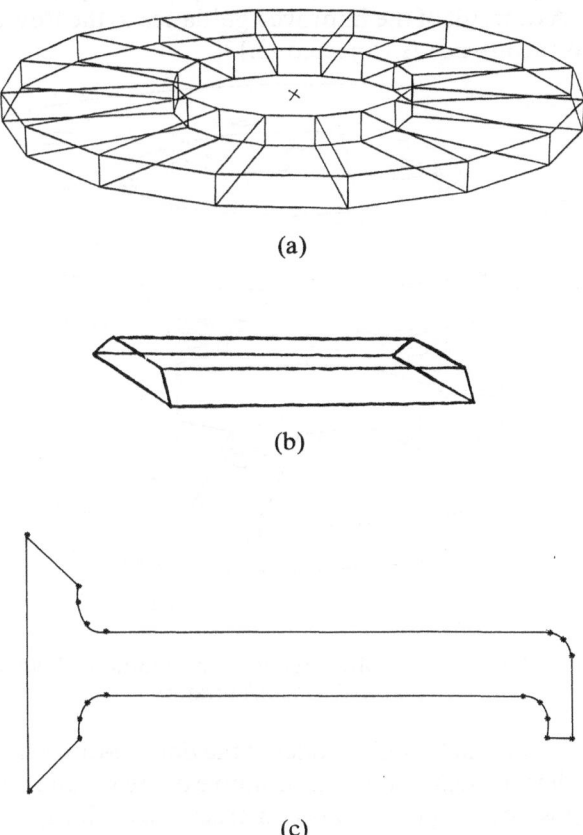

(a)

(b)

(c)

FIG. 3.4. Dove-tailed construction of single-disc arrangement: (a) a faceted tubular primitive; (b) a dove-tail hexahedron primitive; (c) 2-D profile of blade.

blades (only one side can be utilised), (iv) damage caused by escaping shots to blade clamping arrangement. In view of the above, it is believed that a single-disc design can be utilised in applications requiring relatively small size wheel (<250 mm dia). In this case, it is possible to integrate disc, blades and distributor in one unit during the manufacturing process and thus eliminate both lack of rigidity and possible damage to the clamping arrangement of the blades. In view of the present design constraints, a single-disc design was not pursued further.

As an alternative to the above design, a double-disc eight-bladed relatively rigid wheel arrangement is proposed. The details of the design are given in Fig. 3.5; it contains two discs separated by four positioning spacer bars. As a result of the improved guidance to the flow of media, a reduction in turbulence is expected [3.13].

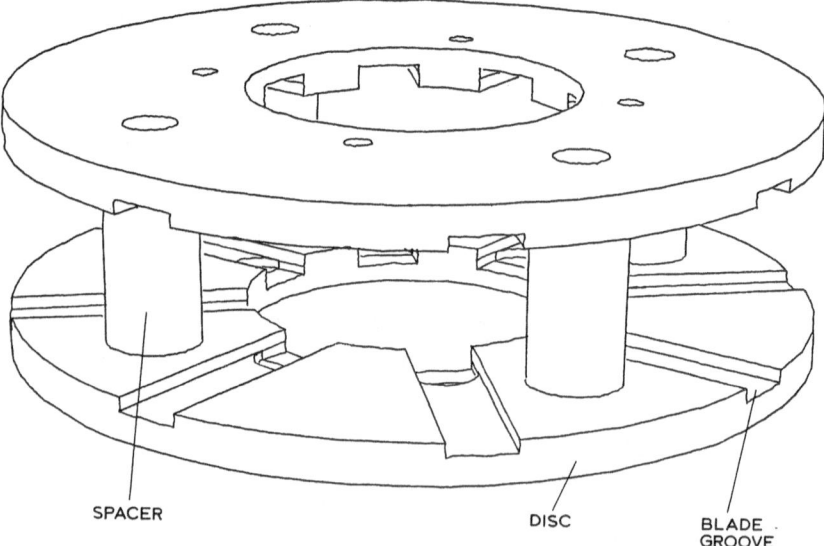

SPACER   DISC   BLADE
                                                       GROOVE

FIG. 3.5. Three-dimensional solid model representation of an eight-bladed double-disc configuration.

In order to construct a solid model of the double-disc arrangement, it was appropriate to create a tubular primitive of the required dimensions. The blade-positioning groove was constructed by joining two blocks of different dimensions to a semi-cylinder, as shown in Fig. 3.6(a). In view of the axisymmetric nature of the disc, the eight positioning grooves of Fig. 3.6(b) were cut out of the disc by rotating it 45° about its axis of symmetry. Four spacer holes were also cut in the formed disc via another

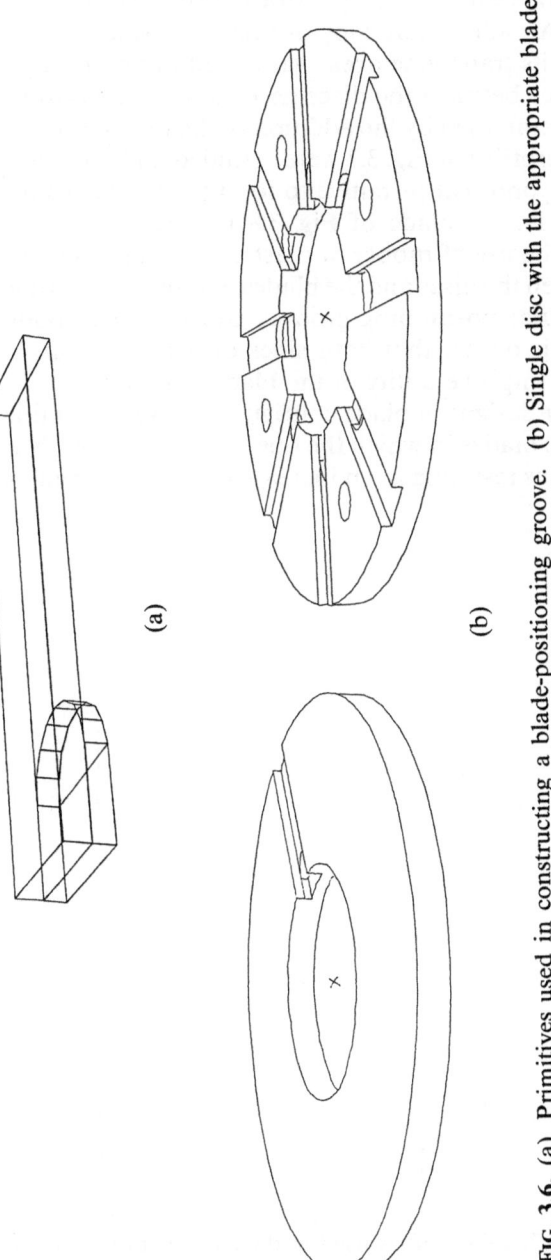

FIG. 3.6. (a) Primitives used in constructing a blade-positioning groove. (b) Single disc with the appropriate blade-positioning grooves.

boolean cut operation of a cylindrical primitive of the appropriate dimensions. An additional copy of this disc, which was rotated 180° about one of its transverse axes, was stored in an appropriate file.

The distance between the discs necessary for the positioning of the blades was maintained by the addition of the four spacers, as shown in Fig. 3.5. The profiles of Fig. 3.7 were extruded and then joined together with a semi-cylindrical primitive to form the blade of Fig. 3.8.

In this study, the blade of Fig. 3.8 is secured in position during standstill by the use of rubber or plastic friction pads which are sandwiched between the discs and the blades. It is secured during rotation by the semi-circular positioning grooves allowed for in both discs. It is clear from this design that both sides of the blade can be used and consequently improve utility of the blade material.

Two further designs of blade configuration were examined. The first utilises curved blades in which the tips are curved towards the direction of rotation, thus resulting in an increase in the impingement velocity of

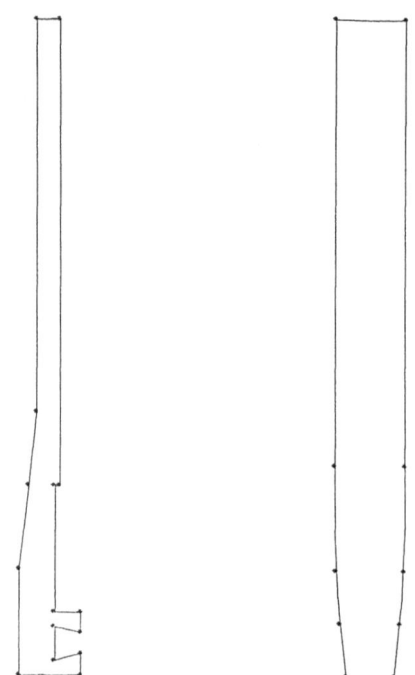

FIG. 3.7. 2-D profiles used in the construction of a blade.

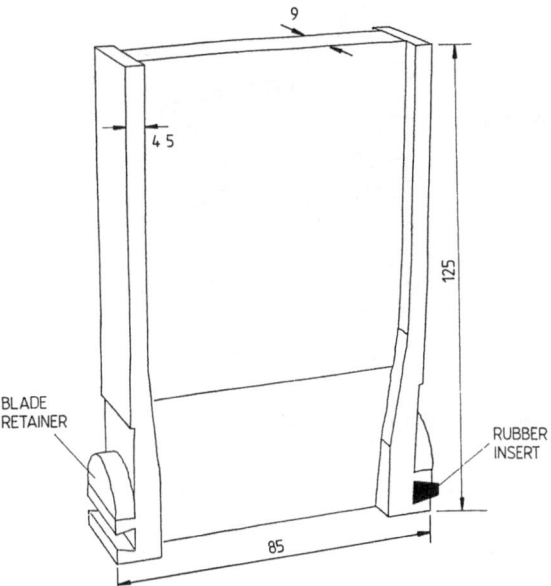

FIG. 3.8. 3-D solid model representation of a blade.

the media in comparison with the straight blade arrangement. This curved type of blade configuration has been used in the single-disc arrangement by some manufacturers. However, it was found that the financial savings resulting from the power reduction are insignificant when compared with the increased production and maintenance costs of both the blades and the supporting disc.

The second design, which is shown in Fig. 3.9, utilises a tubular discharge arrangement [3.14]. In this case, the path taken by the shot is more concentrated than that of a corresponding flat bladed wheel, which results in a higher degree of directional accuracy. Preliminary tests revealed that the resulting peened area is longer and of a much reduced width in comparison with that obtained by the flat bladed wheel arrangement. For this reason, this design alternative was not considered further.

Figure 3.10 shows a solid model representation of an impeller which was mostly constructed from a 2-D profile. The outline profile of Fig. 3.11 of the impeller was defined in the 2-D working space of SDRC-GEOMOD, which is independent of the 3-D work space. The profile

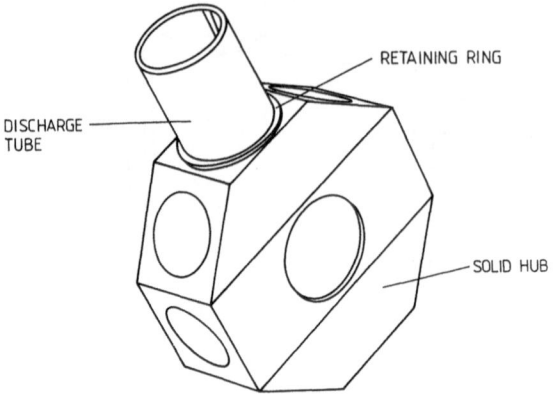

FIG. 3.9. 3-D solid model representation of tubular discharge arrangement.

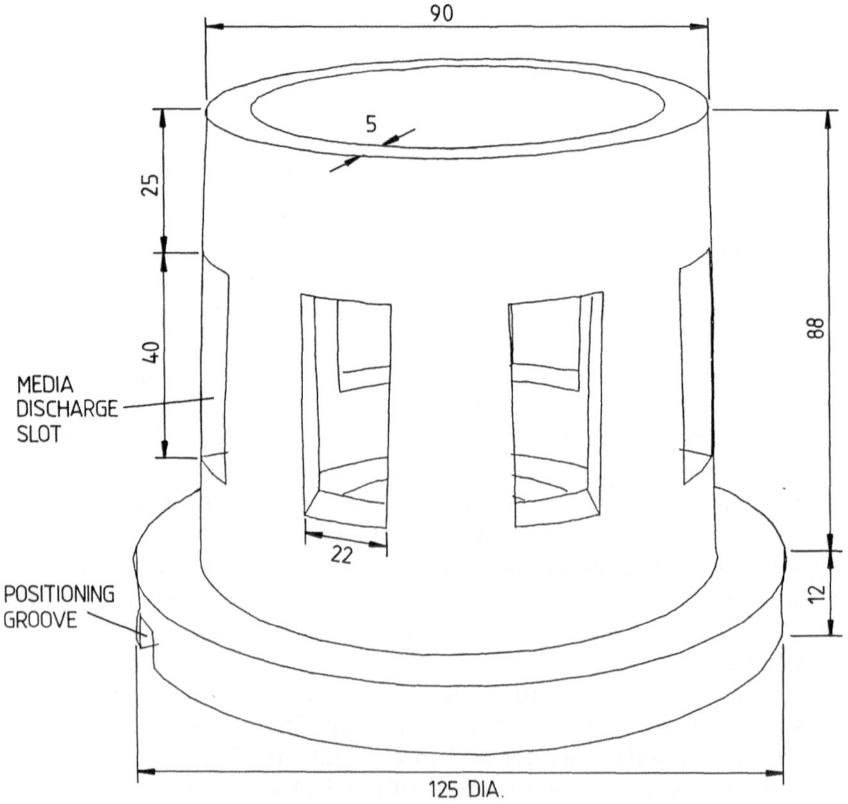

FIG. 3.10. 3-D solid model representation of an impeller.

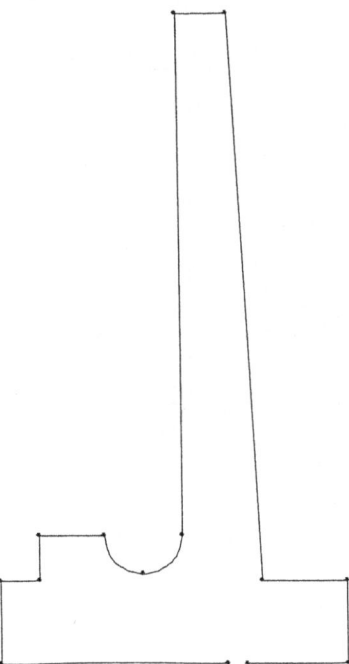

FIG. 3.11. 2-D profile used in constructing the impeller.

was then transferred to the 3-D space to create the main body of the impeller, as shown in Fig. 3.12. The desired slots for media flow were then cut using an appropriate hexahedron primitive. A similar approach was adopted in the construction of the rest of the components of the wheel-assembly. The complex geometrical features of the impeller are clearly demonstrated by the cross-section view of Fig. 3.13.

In this double-disc design, the media is fed by gravity through a feed spout to the core of a matched impeller fastened to the drive shaft. As the media enters the centre of the impeller, it is rotated around and is subjected to centrifugal forces. These forces accelerate the media through the appropriate slot in the impeller into the opening of the control cage of Fig. 3.14. An appropriate clearance between the rotating impeller and the stationary cage is provided to retain the rest of the media within the impeller [3.13].

The angular relationship between the impeller and blades is essential for the accurate delivery of the media to the front of the rotating blades. In this case, a lead angle of 6–8° of the slots in the impeller ahead of the

Integrated Computer-Aided Design of Mechanical Systems

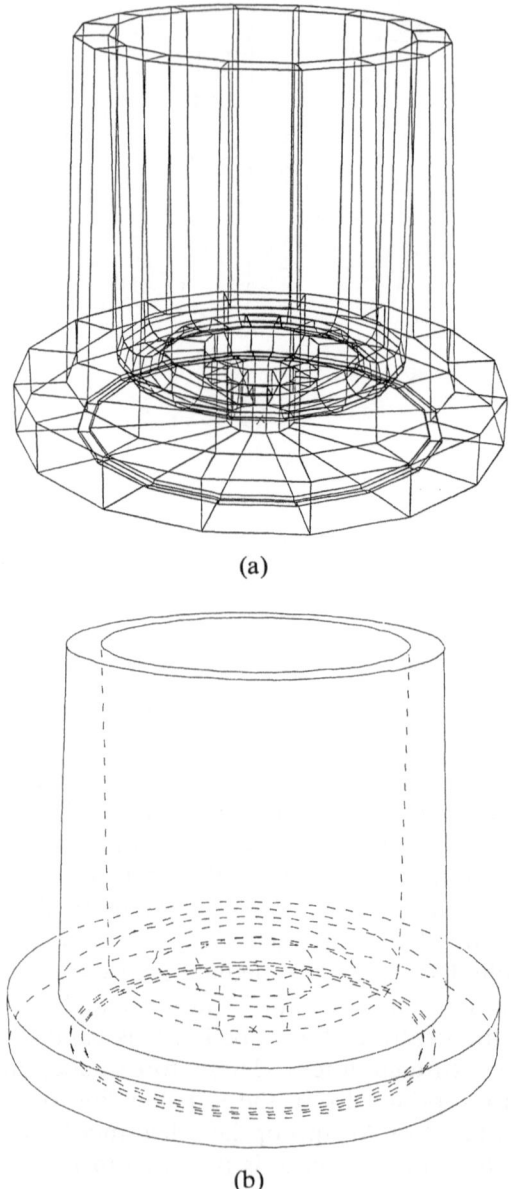

(a)

(b)

FIG. 3.12. Main body of impeller excluding media flow slots: (a) faceted wire-frame representation; (b) accurate representation showing hidden lines dotted.

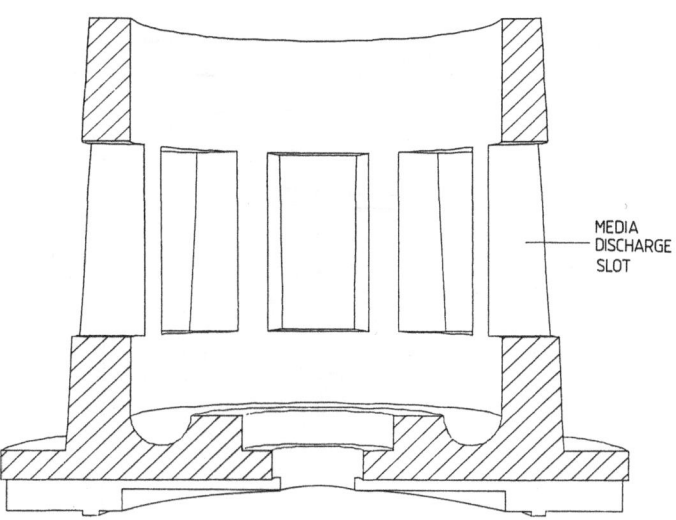

MEDIA
DISCHARGE
SLOT

FIG. 3.13. Cross-sectional representation of the impeller.

blades is recommended. It is also worth pointing out that in order to prevent erratic feeding of media, the number of slots in the impeller was matched to that of the blades.

It is clear from the above that the effect of worn impellers would be to throw abrasives at the trailing edge of the blade or behind it. This disrupts the flow pattern of the media and may cause undue wear upon the back of the blade and/or wheel spacer.

The purpose of the control cage in this design is to provide directional control of the media. The position of the slot in relation to the periphery of the wheel determines both the direction of the discharged media and the region of highly concentrated impingement (known in the industry as the 'hot-spot'). Obviously, the slot shape influences the velocity profile of the impinging shots.

The choice of the appropriate angular velocity of the bladed wheel and the impeller is dictated by the need to produce a suitable impact velocity capable of inducing localised plasticity and maintaining the surface integrity of the exposed target materials.

The complete and accurate determination of the appropriate velocity (taking into account multiple impact effects, statistical nature of the process, strain-hardening and strain-rate effects, surface roughness, etc.) is very complex and the detailed analysis of the problem was considered

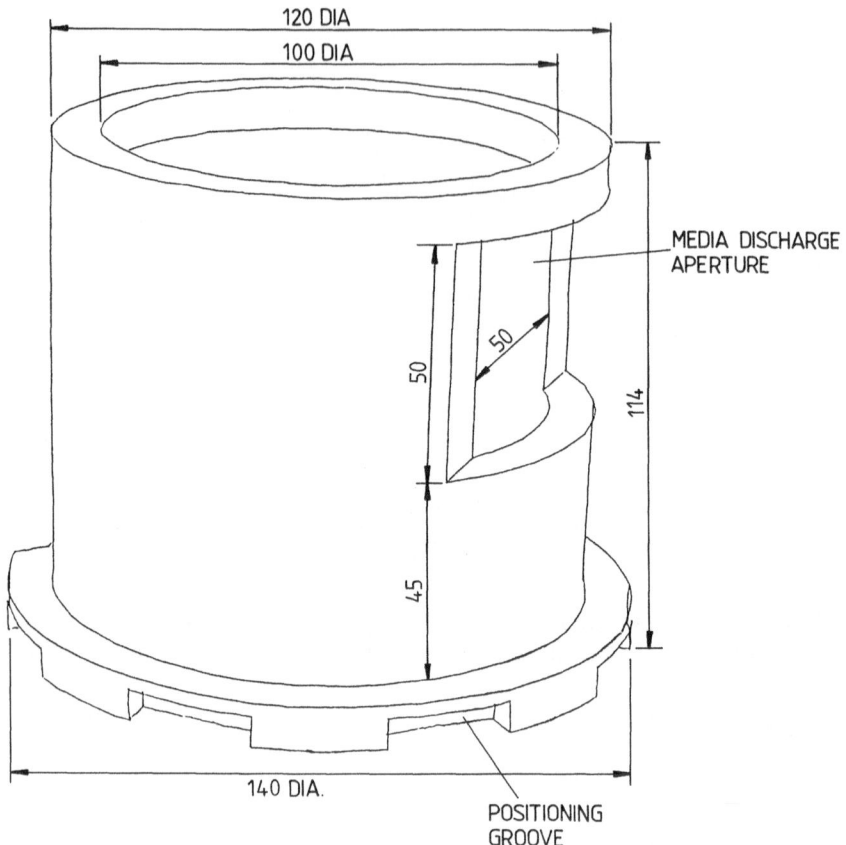

FIG. 3.14. 3-D solid model representation of a control cage.

to be beyond the scope of the present design study. A simplified yet realistic approach was adopted, the details of which are provided below.

3.5.1 A simple model to determine the rotational speed of the wheel

In order to commence the design process, an initial estimate of the impact velocity was necessary for the peening of a range of materials provided in the design constraints. Estimates were based on the following simplified approach for a single shot indenting a target as shown in Fig. 3.15. In this approach, an energy balance relating the net kinetic

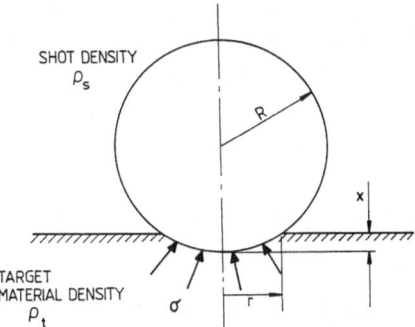

SHOT DENSITY
ρ_s

R

x

TARGET
MATERIAL DENSITY
ρ_t

σ

r

FIG. 3.15. Schematic of a single shot indenting a target.

energy of the impinging shot to the plastic work was invoked. Let us assume that the work done by the indenting shot is given by W:

$$W = \int \sigma A \; dx \tag{3.1}$$

where σ is the resisting pressure of the material ($\simeq 3Y_t$, Y_t the target yield strength). The above expression with appropriate substitutions reduces to

$$W = \int 3Y_t \, \pi r^2 dx \tag{3.2}$$

where r is the maximum radius of the contact area and x is the depth of indentation. Now since $2Rx \simeq r^2$, the above expression becomes

$$W \simeq 3\pi Y_t Rx^2 \tag{3.3}$$

If we assume that the plastic work done by the indenting shot (W) is entirely supplied by the total kinetic energy of that shot ($1/2 \, m v^2$), then

$$\tfrac{1}{2}\left(\tfrac{4}{3}\pi R^3 \rho_s\right) v^2 = 3\pi R Y_t x^2 \tag{3.4}$$

Microstructure analysis of the peened surface as well as surface roughness limitations indicate that a reasonable estimate for depth of distortion is $x = R/10$. The above expression now takes the following form:

$$v \simeq \frac{3}{\sqrt{2}}\left(\frac{R}{10R}\right)\sqrt{\frac{Y_t}{\rho_s}} \tag{3.5}$$

where R is the shot radius, ρ_s its density and Y_t is the target yield strength. Considering a possible increase of 20% in net impact velocity as a result of rebound, the above expression reduces to

$$v \simeq 0.25 \sqrt{\left(\frac{Y_t}{\rho_s}\right)} \qquad (3.6)$$

This enabled the determination of the impact velocity requirement for the different targets as shown in Table 3.1; detailed calculations are given elsewhere [3.6].

Both the dynamics of the present non-rigid system and extensive experimental work on single-bladed wheels reveal that the sliding speed is comparable to the normal speed. In this study, the two speeds were taken to be equal. As a result of this approximation and the assumed geometry of the bladed wheel, an angular velocity of 3000 rev/min was obtained. This angular velocity would enable the peening of the proposed target materials.

Table 3.1
Relationship Between Target Yield Strength (Y_t) and Necessary Impact Velocity for a Cast-Steel Shot

Target yield strength Y_t(MN/m²)	Resultant impingement velocity v(m/s)
200	38·3
300	46·9
400	54·2
500	60·6
600	66·3
700	71·7
800	76·6
900	81·2
1000	85·6

3.5.2 Computer-aided assembly of the different components of the wheel design

Having satisfied the first task of designing the individual components of the wheel-assembly, it is now necessary to investigate the relationship between these components in order to satisfy the present design criteria.

In the present system, the different solid models developed earlier using the Object Modelling module of the package were transferred to

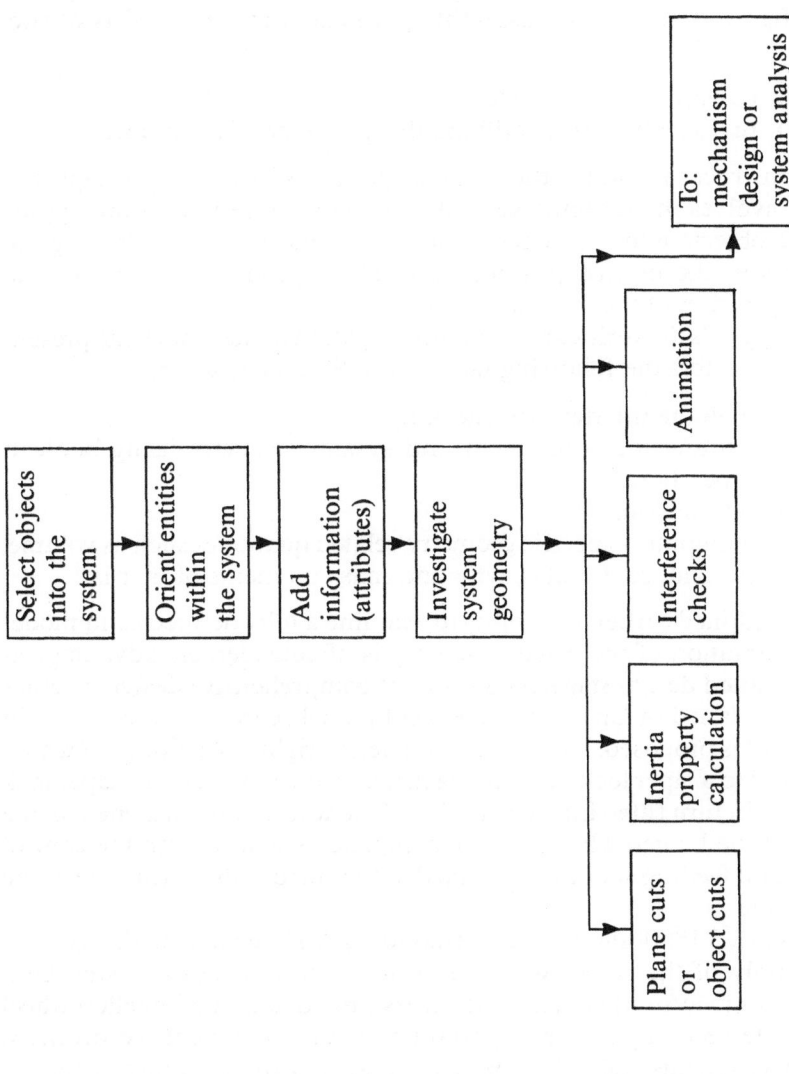

Fig. 3.16. Detailed System Assembly activities of SDRC-GEOMOD (Ref. [3.15]).

the System Assembly module using the same data base. System Assembly allows the designer to select his objects in a logical manner and enables the verification of the system model. Figure 3.16 shows the system modelling activities of System Assembly of GEOMOD [3.15]. The figure indicates that assembling a system model consists of two main activities:

(i) constructing the model,
(ii) investigating and verifying the geometry of the model.

To construct the model, the designer performs two separate steps; the first involves the assemblage of the model via selection of the appropriate objects into the system, and the second involves orienting the objects in the desired position in order to produce the appropriate configuration of the system model.

As regards the verification and investigation of the model, the present package offers the following facilities in System Assembly:

(i)   volume interference checking,
(ii) cutting the component and system geometry using boolean operations,
(iii) animation,
(iv) computing specific geometry/related quantities such as volume, weight, centre of gravity and mass moments of inertia.

As mentioned earlier, the ability to examine a fully defined solid model representation of the wheel-assembly is of considerable advantage at the detailed design stage. As a result of comprehensive design calculations (detailed in Chapter 5), the solid model representation shown in Fig. 3.17 is proposed. It contains, from left to right, the drive shaft which is keyed to a taper lock bush and fastened to the impeller. The taper lock bush is in turn fitted into a wheel hub. The wheel hub is fastened to one side of the bladed wheels. In the complete assembly, both the control cage and feeding spout will be fixed to the surrounding structure of the machine.

Figure 3.18(a) shows a corresponding 2-D section of the general assembly of the wheel with the appropriate dimensions using Unigraphics (McAuto) system. The recess provided at the impeller–wheel hub intersection plays an important role in its alignment. Accordingly, a good fit in this region is expected. It is also worth pointing out that the presence of a groove at the bottom surface of the impeller allows its accurate positioning and prevents any possible relative rotation between

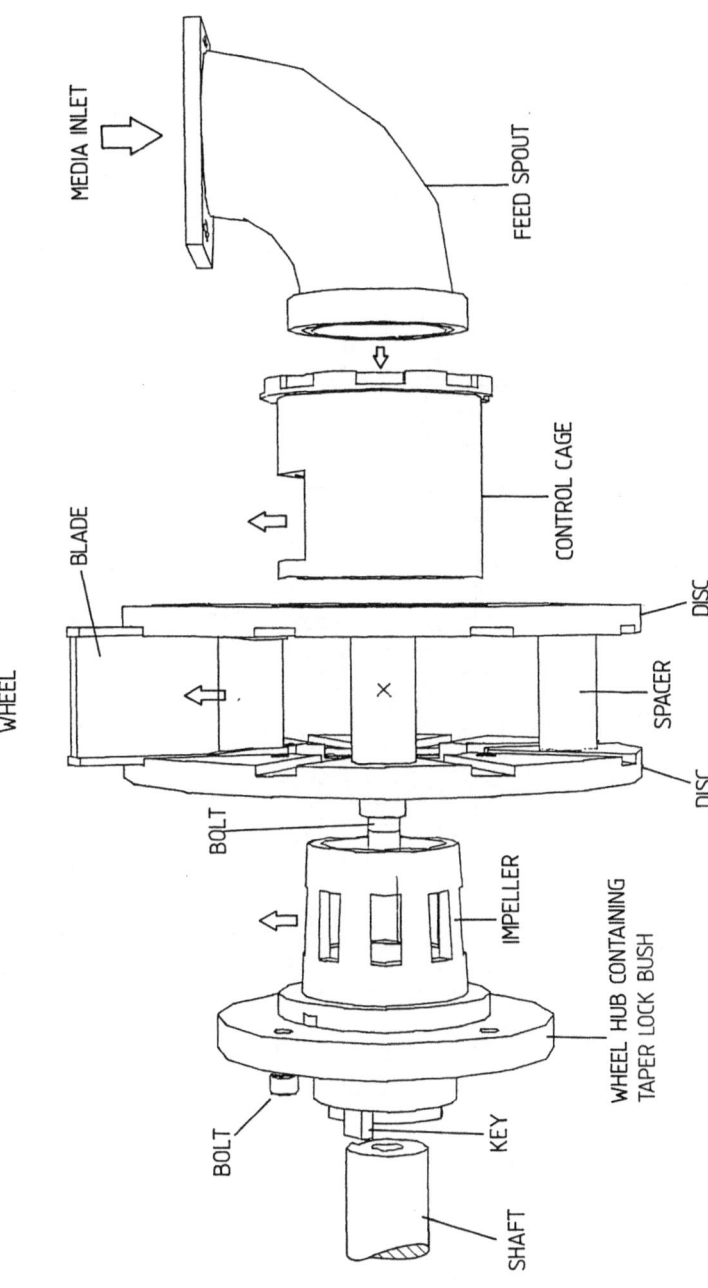

FIG. 3.17. 3-D exploded view of the wheel-assembly using SDRC System Assembly.

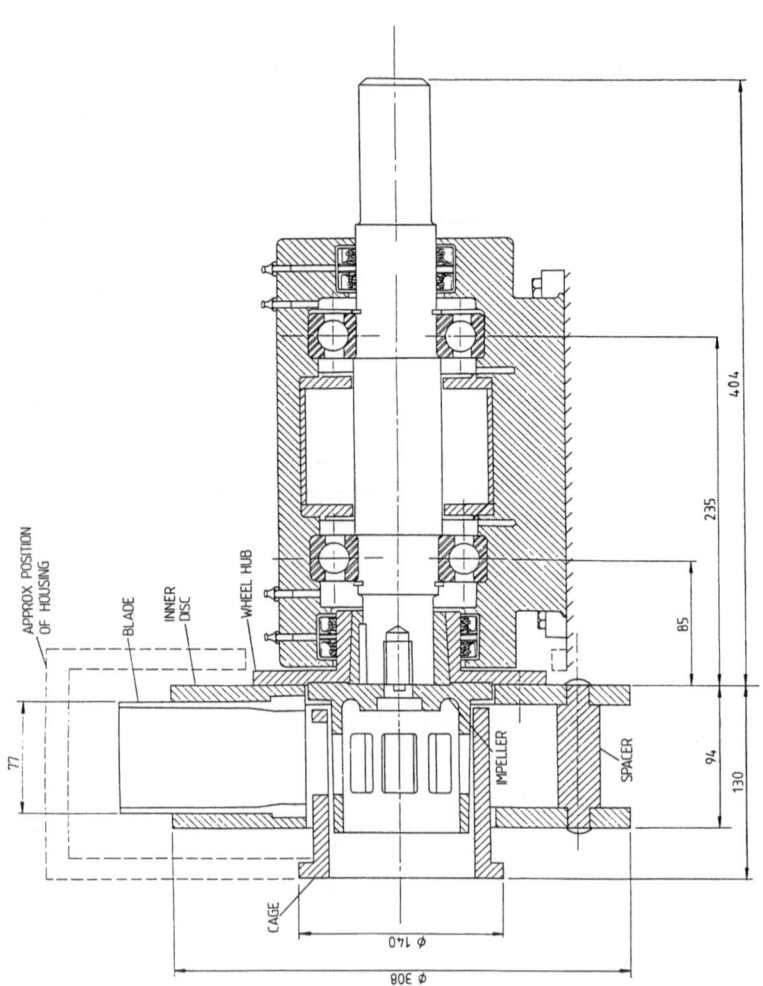

Fig. 3.18. (a) 2-D view of the wheel-assembly using McAuto Unigraphics draughting system (excluding details of input drive system).

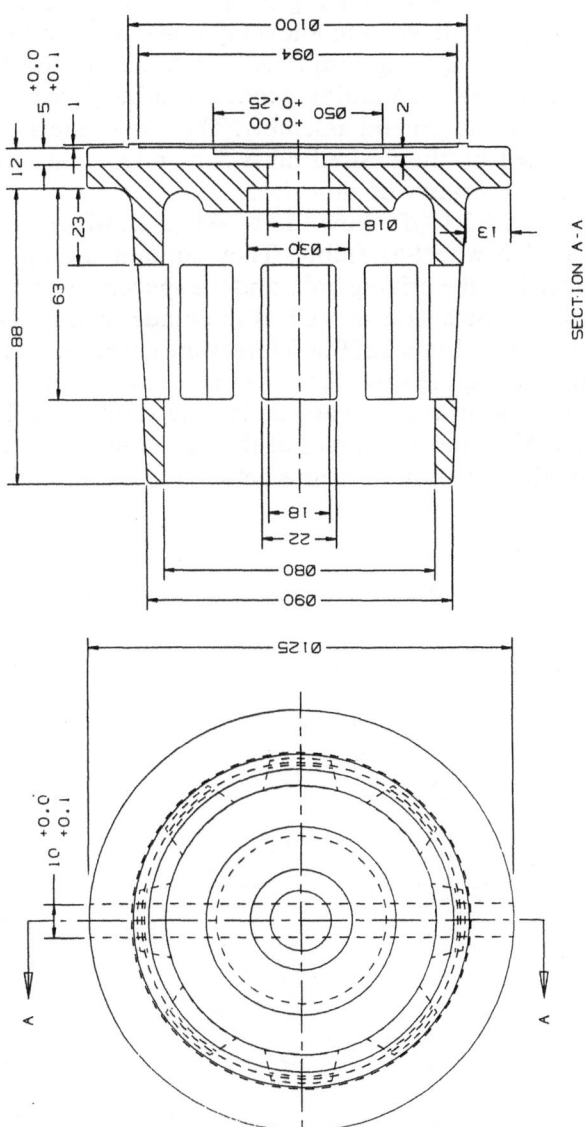

Fig. **3.18.** — *contd.* (b) Detailed working drawing of impeller.

it and the drive shaft during torque transmission. A detailed working drawing of the impeller is shown in Fig. 3.18(b). The impeller is fixed to the drive shaft by one bolt, so as to allow rapid assembly, disassembly and maintenance. The driving torque is transmitted to the impeller from the wheel hub via a longitudinal groove running perpendicularly to the view shown in the general assembly. The size of the slots in the impeller is governed by the desired mass flow rate capability of the equipment for a given range of shot size.

Figures 3.18(c) and 3.18(d) show the working drawings of the two rotating discs of the wheel-assembly. They contain the appropriate holes to accommodate the driving bolts and the spacers. Both discs also show the radial guidance slots as well as the blade retainer grooves.

The size and shape of blades of Fig. 3.18(e) were governed by the mass flow rate capability and the structural integrity of the system as well as the geometrical constraints of the assembly. The surface profile of the blade was dictated by the kinematics of the travelling shots, and thus ensure the smooth transition of media between impeller and blade.

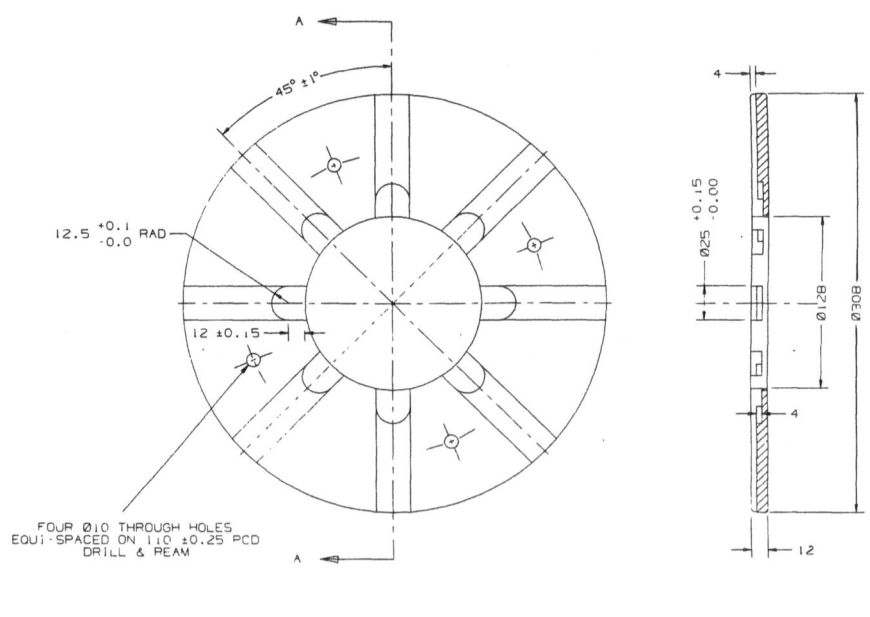

FIG. 3.18. — *contd.* (c) Detailed working drawing of outer-disc.

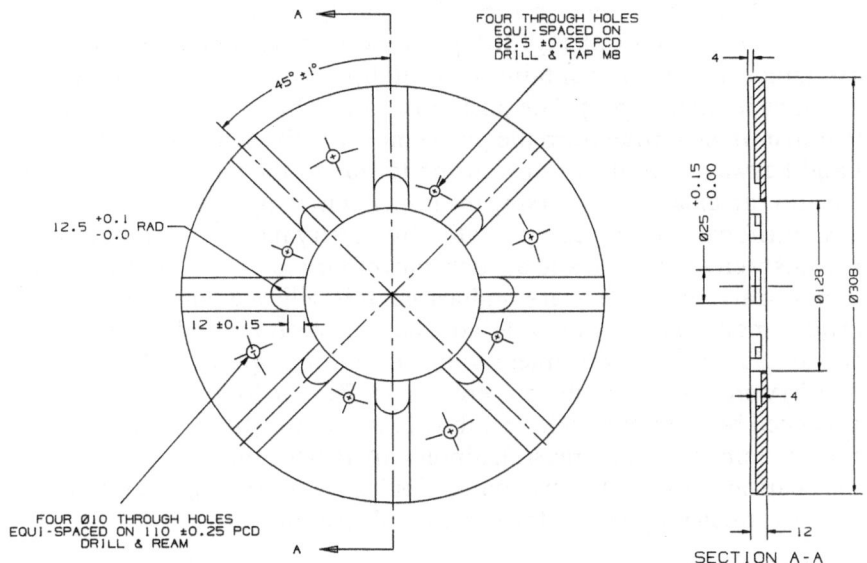

FIG. 3.18. — *contd.* (d) Detailed working drawing of inner-disc.

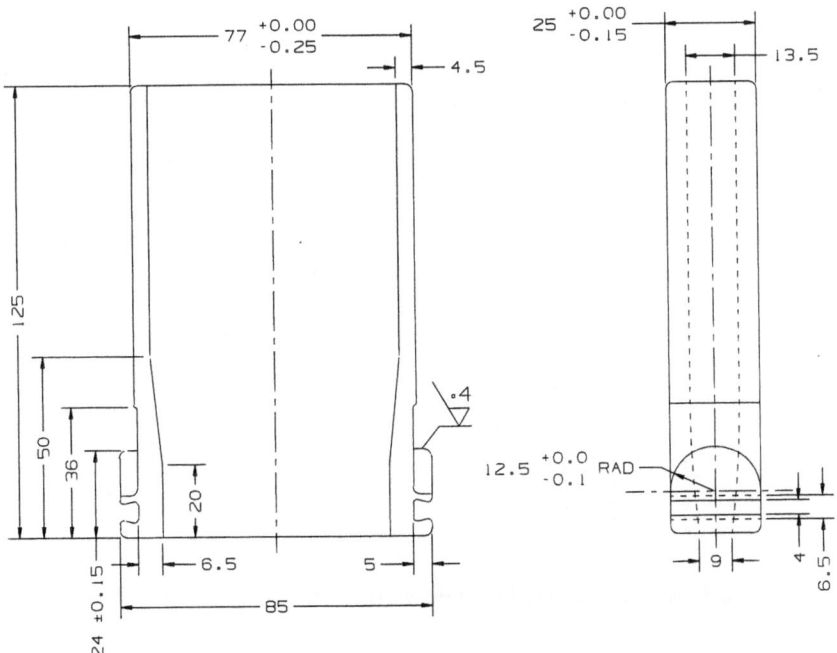

FIG. 3.18. — *contd.* (e) Detailed working drawing of blade.

The wheel hub shown in Fig. 3.18(f) is used to transmit the torque from the input shaft of Fig. 3.18(g) to the impeller and to the wheel-assembly. The shaft contains two circlip-grooves for retaining the bearings in their appropriate positions. The right-hand keyway is used to transmit the torque from the prime mover to the shaft, while the left-hand keyway is used to transmit the torque to the wheel hub. It also contains an axial hole for fixing the impeller in its appropriate position.

Centrifugal peening equipment is, by its very nature, bound to work in a dusty environment, and as such, the importance of the correct selection of seals cannot be over-emphasised. It is believed that twin-lip seal arrangement may not provide adequate protection, and accordingly two single-lip seals with integral dust excluding lips as shown in Fig. 3.18(h) were adopted in the present design. For added protection, grease is packed between the seals and thus prevents dust from damaging the rolling elements. Four greasing nipples are located on the top section of the bearing housing. The two outer nipples are for sealing grease, while the two inner nipples are for bearing lubrication.

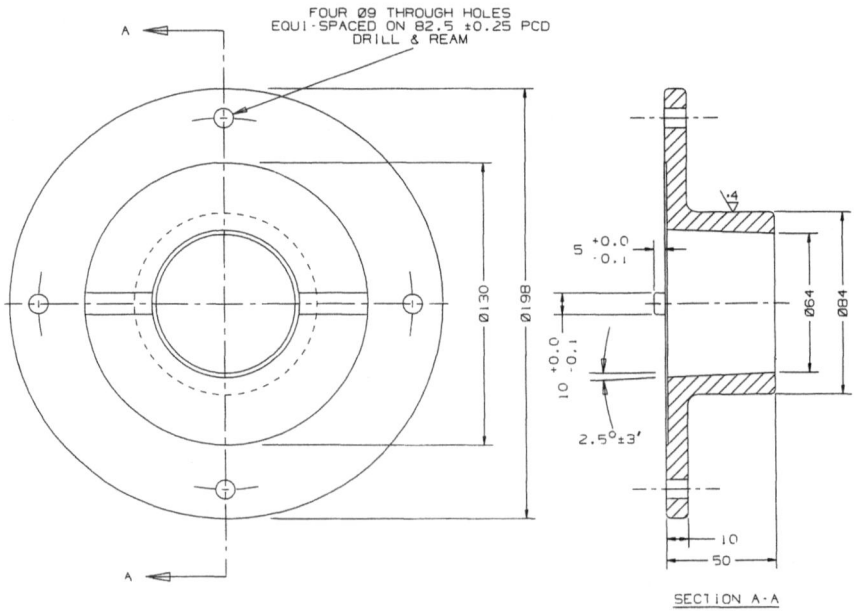

FIG. 3.18. — *contd.* (f) Detailed working drawing of wheel-hub.

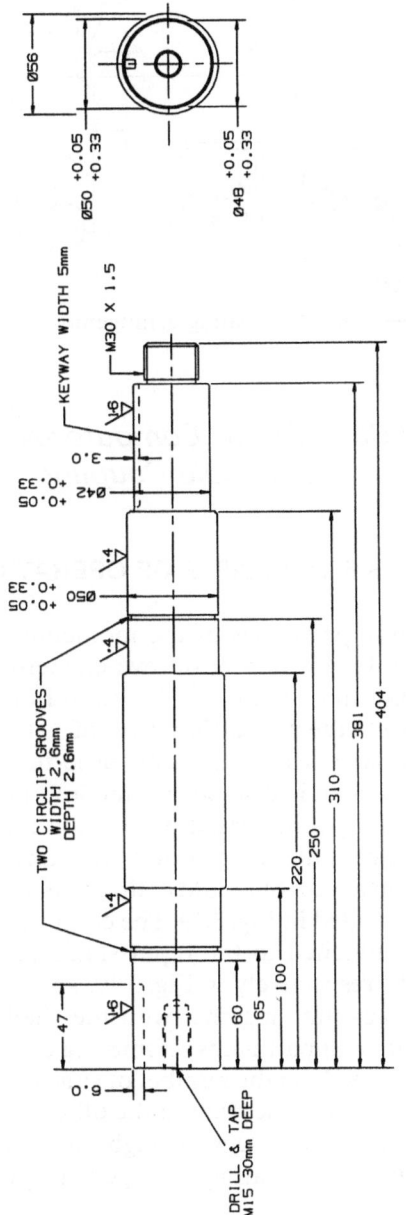

Fig. 3.18. — *contd.* (g) Detailed working drawing of drive shaft.

FIG. 3.18. — *contd.* (h) Sealing arrangement used in the design.

Second Case Study: Computer-Aided Design of a Fluid Coupling

3.6 PRINCIPLE OF OPERATION

The hydrodynamic principles of the momentum transfer and power transmission employed in a basic two-element fluid coupling were developed by a German engineer, Hermann Föttenger, in 1908. It was originally applied to the propulsion drive of a ship, but since then hydraulic couplings have also been widely used in mining, iron and steel production, petro-chemical, food production, automotive and power generation drives. With increased production and wider applications, great interest is currently being shown by many engineers.

The basic construction of a fluid coupling is shown by the schematic cross-section illustrated in Fig. 3.19. The combined impeller and coupling housing are attached to the input shaft. The impeller and rotor, labelled I and R* respectively in Fig. 3.19, are similar in construction. Each has flat vanes that, together with the shell and core, divide the space within it into compartments resembling distorted half 'doughnuts'. For reasons discussed later the number of vanes in one of the elements is usually about 0·90–0·95 times that in the other. The rotor is attached to the output shaft which emerges through an opening in the housing.

The introduction of a sealing arrangement prevents leakage of the

---

*It is possible to interchange the terminology of impeller and rotor in some engineering applications.

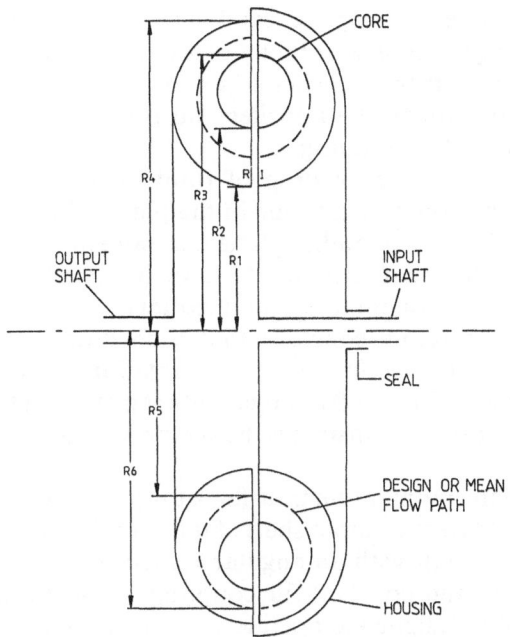

FIG. 3.19. Schematic cross-section of a fluid coupling.

working fluid from the housing at the exit of the output shaft. This working fluid is a light oil which largely or completely fills the space within the housing. In some situations it is necessary to continuously remove and return the fluid during operation to pass it through a cooling device which dissipates the fluid heat. A primary cause of this heat is the viscous friction which acts within the fluid.

Assume that the input shaft and impeller are driven by a power source at a constant speed, ω_i rev/min, and that the input shaft is held stationary. Consider a very small volume of fluid in the impeller (a drop of fluid) at the radius R_2 from the axis of the coupling. The drop is rotated around the axis of the coupling at a speed of ω_i rev/min as a result of the force exerted on it by an impeller vane either directly or through other similar drops of fluid. The centrifugal forces move this drop and other drops in the impeller towards the outer part of the impeller and across the small gap into the rotor. Since the rotor is stationary no centrifugal forces are exerted by the drops, due to its rotation, after they enter the rotor. Therefore, the drops flow through the rotor, across the inner gap between the two elements, and back into the impeller. From here the drops repeat the cycle of movement over and over again as long as the speed of rotation of the impeller is greater than that of the rotor. Thus the impeller acts as a centrifugal pump and circulates the fluid through the impeller and rotor in the planes which include the axis of the coupling.

In these movements, each drop of fluid in the impeller has two rotational velocity components: one in the planes that are radial from the axis of rotation, or the radial plane component; the second in the planes that include that axis, or the circulatory plane component. Obviously these components are in planes that have 90° angles between them. One can therefore conclude that each drop of fluid has two velocity components; one for each of the rotational components, and these at any instant have a 90° angle between them. The total linear velocity of any drop at any instant is the vectorial sum of the two tangential velocities.

As a result of the momentum transfer, a torque is exerted on the rotor which tends to rotate the output shaft of the coupling in the same direction as the input shaft with an angular velocity ω_r rev/min.

A variation of the constant-fill coupling is the variable-fill type, which is specially designed to operate at high slip rates. The hydraulic fluid fill level is pre-adjusted to control output speed and slip time. However, this action also produces turbulence and generates heat at high operating speeds, and as a result of this auxiliary coolers are often fitted to maintain safe operating temperatures.

Fluid couplings are the smoothest currently operating slip type coupling for high torque transmission and are thus well suited for soft starts and protection against load shocks and jam-ups and for economical regulation of the output of pumps, fans, blowers, etc. They also dissipate heat well through the fluid fill either to ambient air or auxiliary coolers. Therefore, they are often used when frequent starts and stops are required and where high inertia loads require long acceleration time. Because the coupling bearings are constantly lubricated by the oil fill, fluid couplings have long service lives with little or no maintenance.

3.7 DESIGN CONSTRAINTS AND REQUIREMENTS

The aim of this part of the work is to design, from first principles, a constant filling fluid coupling between an electric motor and a gearbox servicing a high speed centrifugal pump as depicted by the schematic of Fig. 3.20. The pump is to be installed in a nuclear power station and will

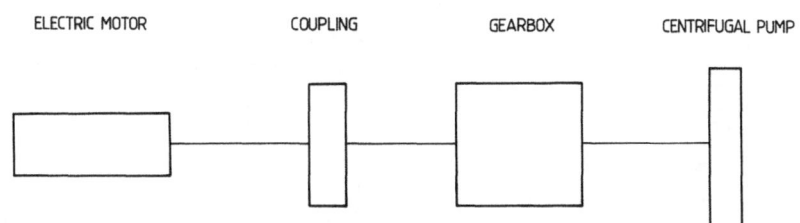

ELECTRIC MOTOR   COUPLING   GEARBOX   CENTRIFUGAL PUMP

FIG. 3.20. Schematic layout of the provided transmission system. (Another alternative will be to place coupling between gearbox and pump.)

only be used in the event of a breakdown in the main boiler feed pump system.

The major design considerations were:

Input power	1·24 MW
Input speed to coupling	1500 rev/min
Pump head	2000 m
Pump flow rate	0·1 m³/s
Output speed from gearbox	6000 rev/min
Maximum spatial radial clearance	0·35 m
Efficiency	95%

The above mentioned design constraints are in addition to the conventional ones of being safe, easy to maintain, reliable and economical to manufacture.

3.8 FLOW MODELS AND ANALYSIS

In order that the flow within the coupling can be analysed, a suitable flow model must be chosen. Two models were considered; the first assumes a uniform velocity distribution as shown in Fig. 3.21(a), while the second assumes a linear distribution as depicted by Fig. 3.21(b). For details see Refs [3.16] and [3.17].

Figure 3.21(a) shows that the uniform velocity distribution is between the inner and outer radii R_1 and R_4, with the fluid rotating about a mean radius R_M. This simple flow model leads to an unrealistic discontinuity in the velocity distribution between the radii R_1 and R_4.

The above two flow models were compared with experimental results obtained from testing a fully filled coupling detailed in [3.16]. The results indicate that acceptable differences exist between the two, and

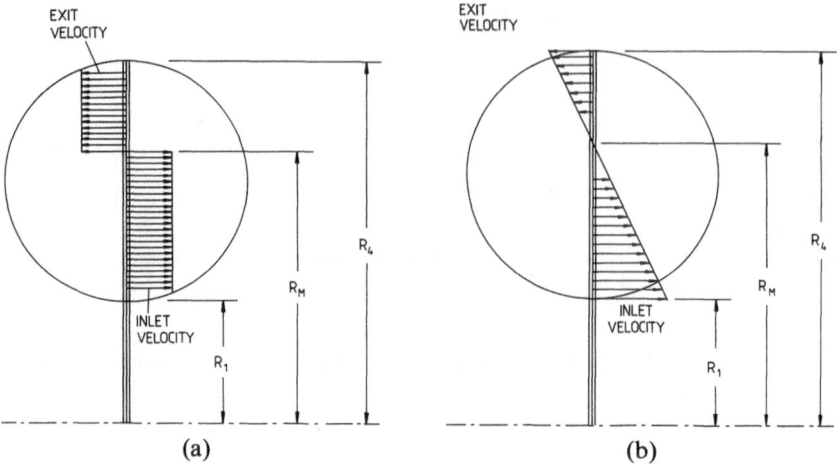

FIG. 3.21. Assumed flow model in the coupling: (a) uniform velocity distribution; (b) linear velocity distribution.

also confirm that both models compare favourably with the experimental results.

Based upon the uniform velocity distribution and continuity condition of equal flow rates in the upper and lower portions of the working circuit, Qualman and Egbert [3.17] developed the following formulae to enable the determination of the characteristic dimensions R_M, R_5 and R_6 as follows:

$$R_M = \sqrt{\frac{R^2_1 + R^2_4}{2}} \tag{3.7}$$

$$R_5 = \sqrt{\frac{R^2_1 + R^2_2}{2}} \tag{3.8}$$

$$R_6 = \sqrt{\frac{R^2_3 + R^2_4}{2}} \tag{3.9}$$

The torque capacity of a coupling depends upon its speed, size, fluid characteristics and internal design. If one assumes that a fluid of density ρ enters and leaves the rotor at the same circulation rate Q, then the net torque T on the rotor is

$$T = \rho Q (R^2_6 \omega_i - R^2_5 \omega_r) \tag{3.10}$$

where R_5 is the internal effective radius, R_6 is the external effective radius for the design of mean flow path, and ω_i and ω_r are the angular rotational velocities of the impeller and rotor, respectively. By assuming a realistic value for the coupling slip rate S, which is defined as

$$S = \frac{\omega_i - \omega_r}{\omega_i} \qquad (3.11)$$

the specific torque (per unit mass flow rate $\dot{m} = \rho Q$) can be determined. The power input to the impeller is therefore

$$P_i = \rho Q \omega_i (R^2_6 \omega_i - R^2_5 \omega_r) \qquad (3.12)$$

and the output power from the rotor is

$$P_r = \rho Q \omega_r (R^2_6 \omega_i - R^2_5 \omega_r) \qquad (3.13)$$

The power dissipated as heat is then given by the difference between equations (3.12) and (3.13), as follows:

$$P_L = P_i - P_r = \rho Q (\omega_i - \omega_r)(R^2_6 \omega_i - R^2_5 \omega_r) \qquad (3.14)$$

From the knowledge of the power required, the torque transmitted and an assumed slip value of 5%, the following estimates for the basic dimensions of the coupling were obtained using the appropriate mass flow rate equations (see Refs [3.18] and [3.19]):

$$R_1 = 0{\cdot}01 \text{ m}$$
$$R_2 = 0{\cdot}18 \text{ m}$$
$$R_3 = 0{\cdot}24 \text{ m}$$
$$R_4 = 0{\cdot}30 \text{ m}$$

The value of R_4 was dictated by the provided design constraints.

If the efficiency of the fluid coupling is assumed to be 95%, then the power loss in the form of heat generation is approximately 60 kW. This heat must be dissipated to coolers, otherwise the working fluid of the coupling will overheat and its characteristic will deteriorate.

3.9 CONCEPTUAL DESIGN

Figure 3.22(a) shows a 2-D section view of the general assembly with appropriate dimensions using McAuto Unigraphics system. It contains from right to left: the input shaft, inlet and outlet cooling system arrangement, detailed in Fig. 3.22(b), the bearing cover which contains

mechanical seals, the housing, the impeller, the rotor which is effectively the output shaft and finally the bearing support. The assembly
shows that the impeller and the rotor utilise semi-circular vanes which
together with the inner shells divide the space within it into fluid compartments. The present design allows for the use of a core of space
having no vanes which is filled with fluid. A function of the core is to
reduce circulatory flows of fluid near the centre of the flow paths.

The above 2-D modelling approach was supplemented by the use of
solid models to aid its visualisation and to perform volume interference
checks. Figures 3.23 and 3.24 show the different design aspects of both
rotor and impeller. In this case, the hybrid solid modelling approach
was adopted. The main body of both rotor and impeller was developed
using the 2-D profile representation of SDRC-GEOMOD, as shown in
Fig. 3.25. These profiles were then transferred to the 3-D object modelling section of the package and revolved 360° about the appropriate axis

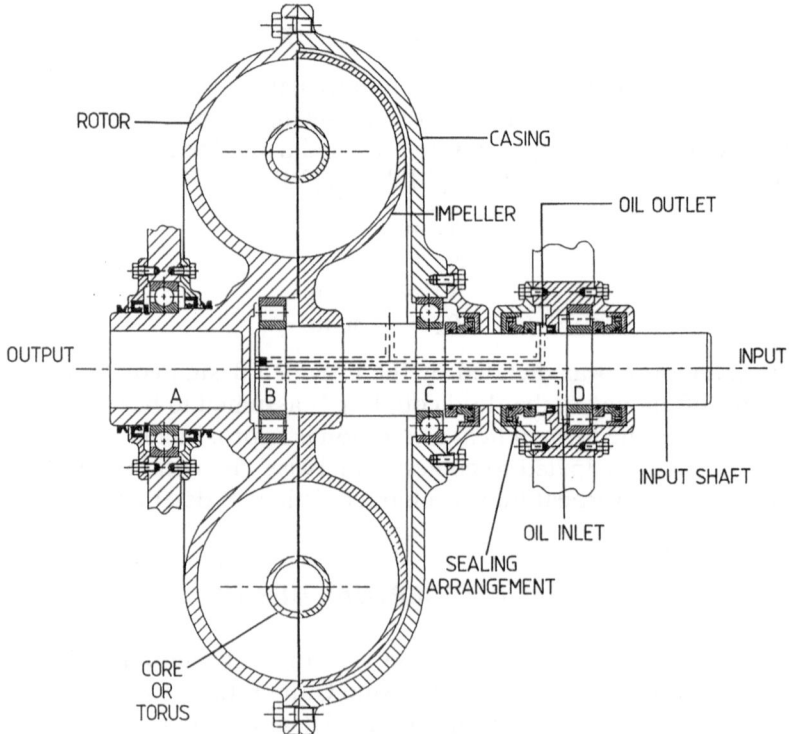

FIG. 3.22. 2-D layout of the coupling assembly using the McAuto Unigraphics
system. (a) General assembly.

to generate the necessary shells. The necessary boolean operations were then performed on the geometries of Fig. 3.26 in order to construct the solid models of both rotor and impeller. In this case, the sequence was to join the vanes to the shells and this was then followed by joining the core to the vanes, thus making complete integrated components. The bolt holes in the rotor were cut using a cylindrical primitive tool via a boolean cut operation.

In the present design, the number of vanes in the impeller was taken to be 36, while in the rotor it was 33. This number was dictated by the need to:

   (i) transmit a given torque T at a desired angular velocity ω,
  (ii) reduction of eddy flows or localised vortices in the impeller and reversed flows in the rotor,
 (iii) structural integrity,
 (iv) clearer channels for the flow of fluid in the circulatory planes, and
  (v) reduction of noise or 'hum' resulting from the coupling in comparison with that produced from a coupling with equal numbers of vanes in the elements.

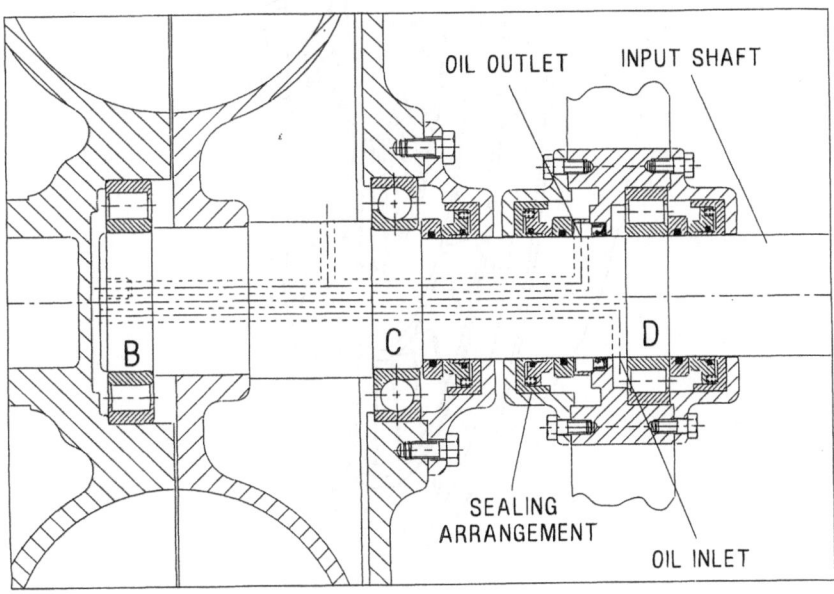

FIG. 3.22. — *contd.* (b) Enlarged section of the assembly showing bearing arrangement, oil seals and cooling oil gland.

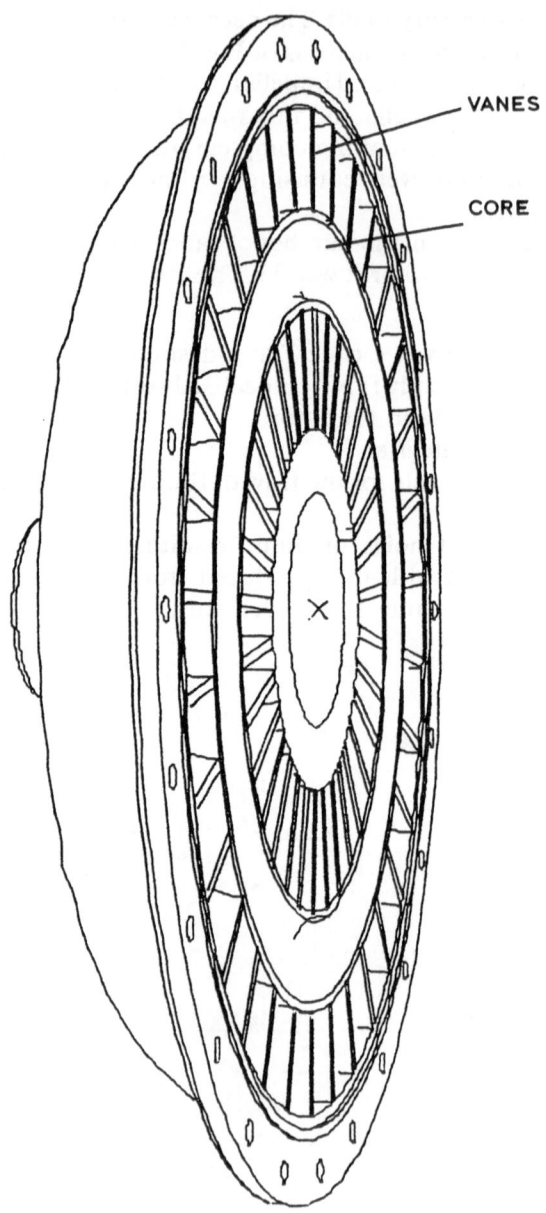

FIG. 3.23. (a) Solid model view of rotor showing vanes, core and fixing holes.

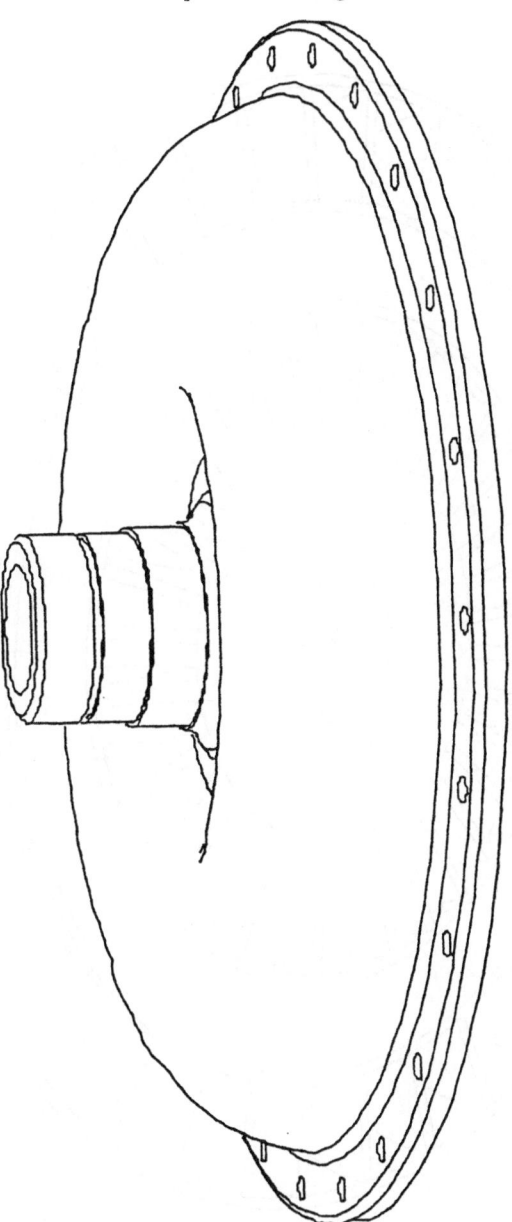

FIG. 3.23.—*contd.* (b) Solid model view showing the back of the rotor.

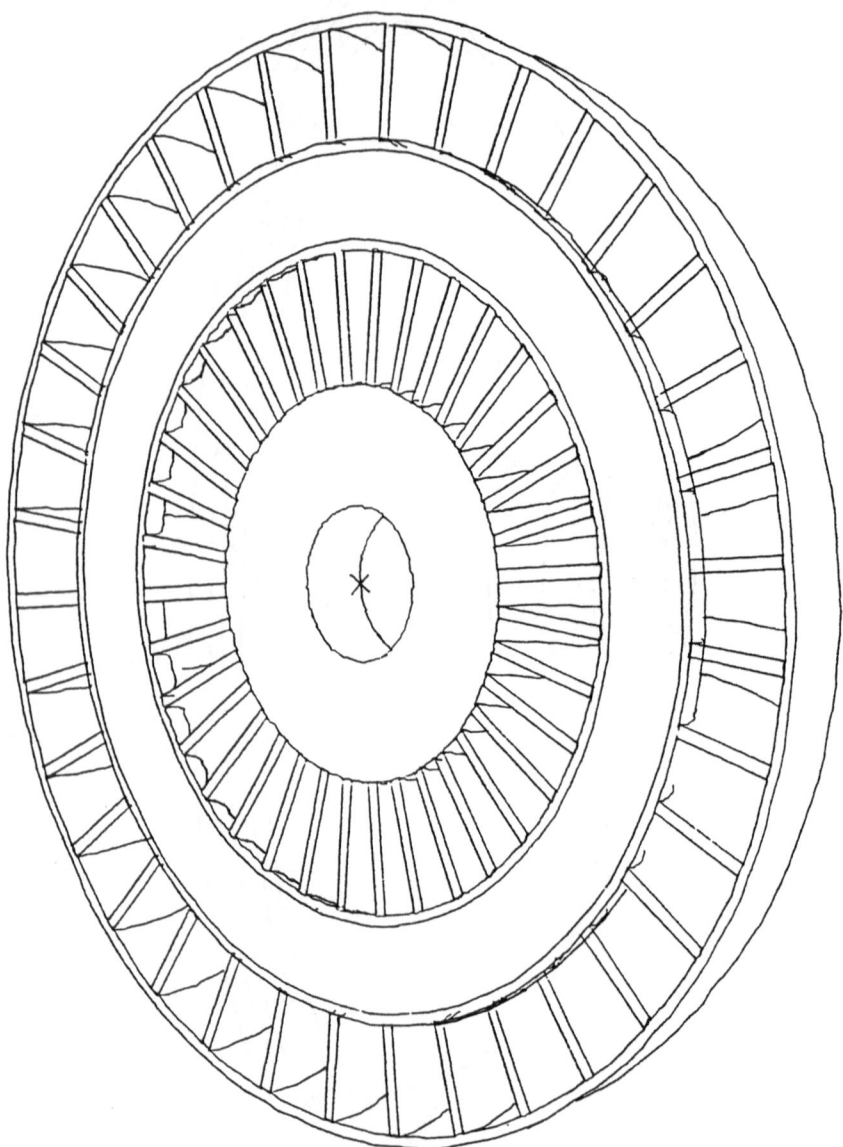

FIG. 3.24. Solid model view of impeller with vanes and core attached.

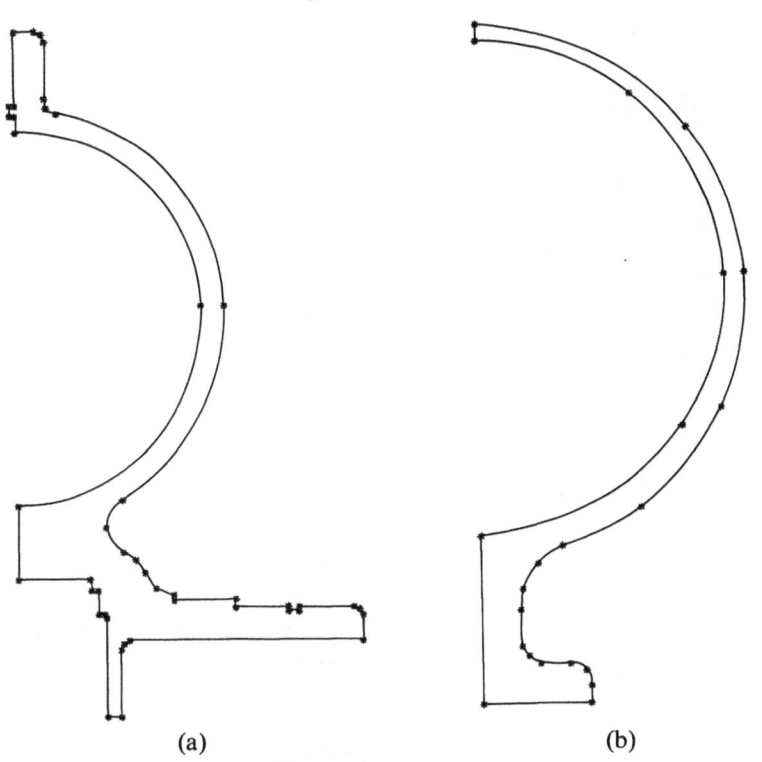

(a)   (b)

FIG. 3.25. 2-D profiles used in the construction of (a) rotor; (b) impeller.

A section of the impeller showing the vanes attached to both shell and core is provided in Fig. 3.27.

Similarly, the housing of the coupling input shaft and cover were developed using 2-D profile representation of GEOMOD. Figures 3.28 and 3.29 show the corresponding solid models of the two components.

The ability to examine a fully defined solid model representation of the different components of the coupling is of considerable value to the detailed design stage. As discussed earlier, this enables the verification of the relationship between the different components of the model assembly. Figure 3.30 shows an exploded solid model representation of the assembly, while Fig. 3.31 shows the corresponding colour-shaded image which was developed using the system assembly of GEOMOD.

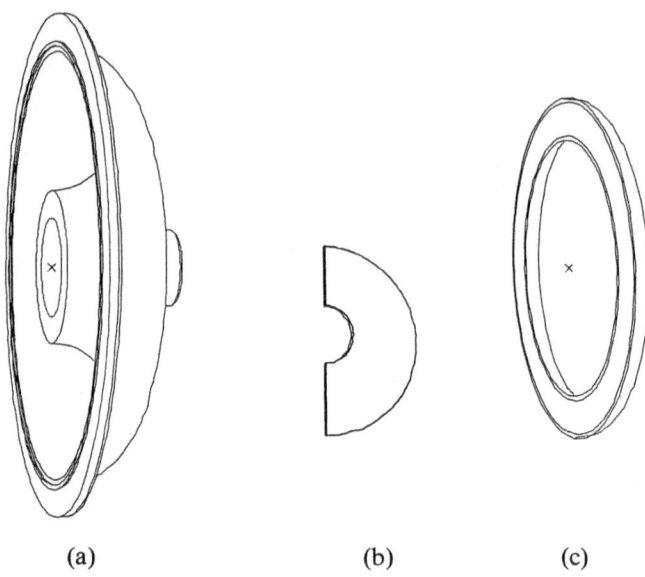

(a)   (b)   (c)

FIG. 3.26. Different solid models used in generating the rotor: (a) the shell; (b) vane; (c) core.

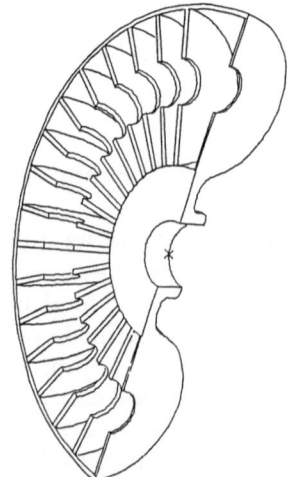

FIG. 3.27. A section of impeller showing vanes and core.

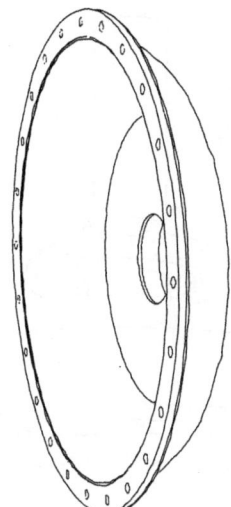

FIG. 3.28. Solid model of casing.

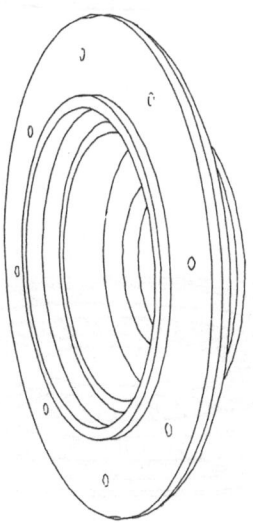

FIG. 3.29. Solid model of cover.

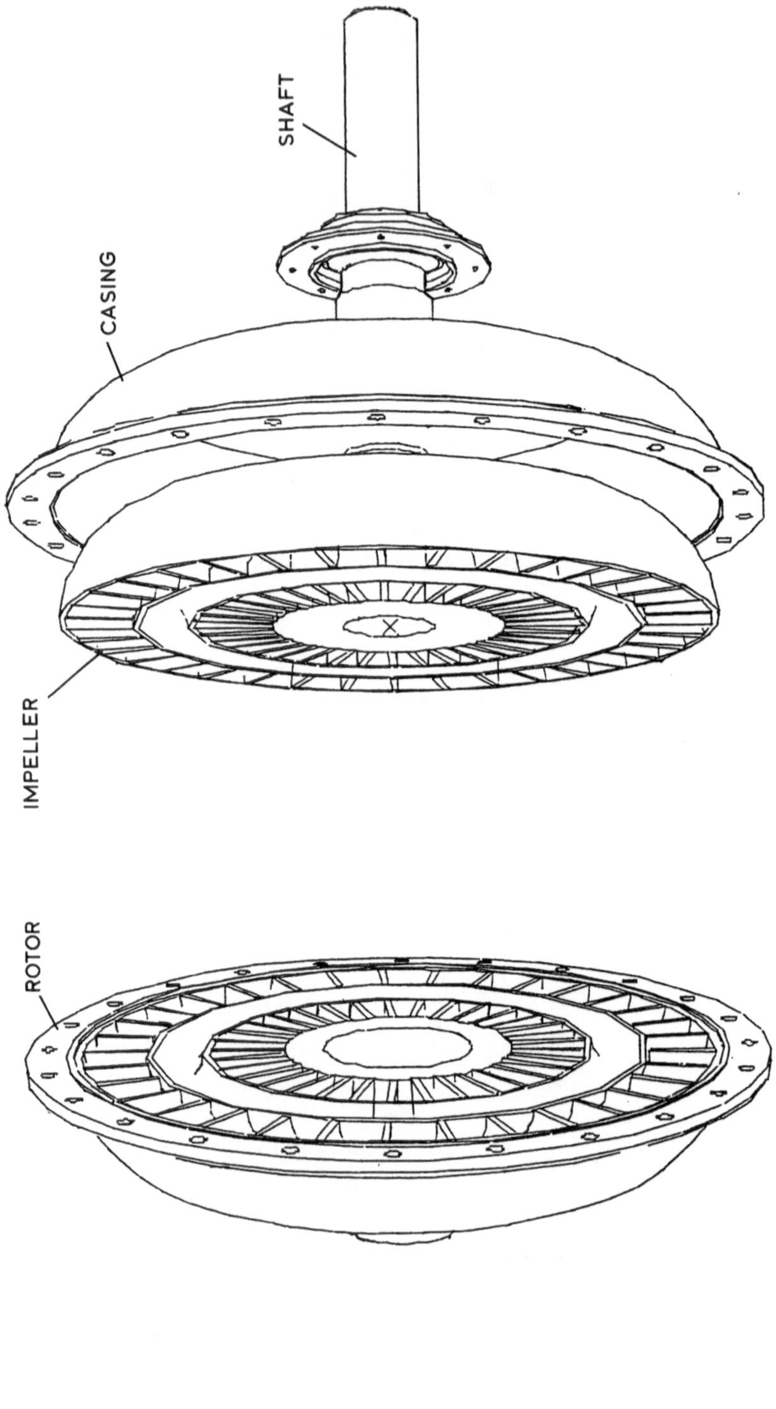

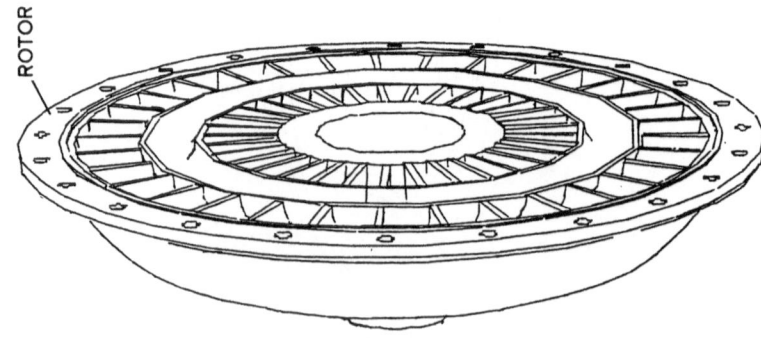

FIG. 3.30. An exploded solid model representation of the coupling assembly.

FIG. 3.31. An exploded solid model of the assembly using the colour-shaded image of SDRC-GEOMOD.

In order to perform volume interference checks, the different components were assembled in their appropriate positions as shown in Fig. 3.32. It is interesting to note that there exists a transparent facility in most packages which allows the visualisation of hidden components within the assembly.

Figure 3.33 shows a typical example of the assembly with the housing made transparent.

From the amount of heat generated within the coupling it was thought necessary to utilise some form of cooling system. One possible method of increasing the rate of cooling is to add fins to both rotor and housing. These fins increase the surface area of the components and thereby increase the rate of heat dissipation. However, the introduction of these fins may not be adequate for this present application. A more

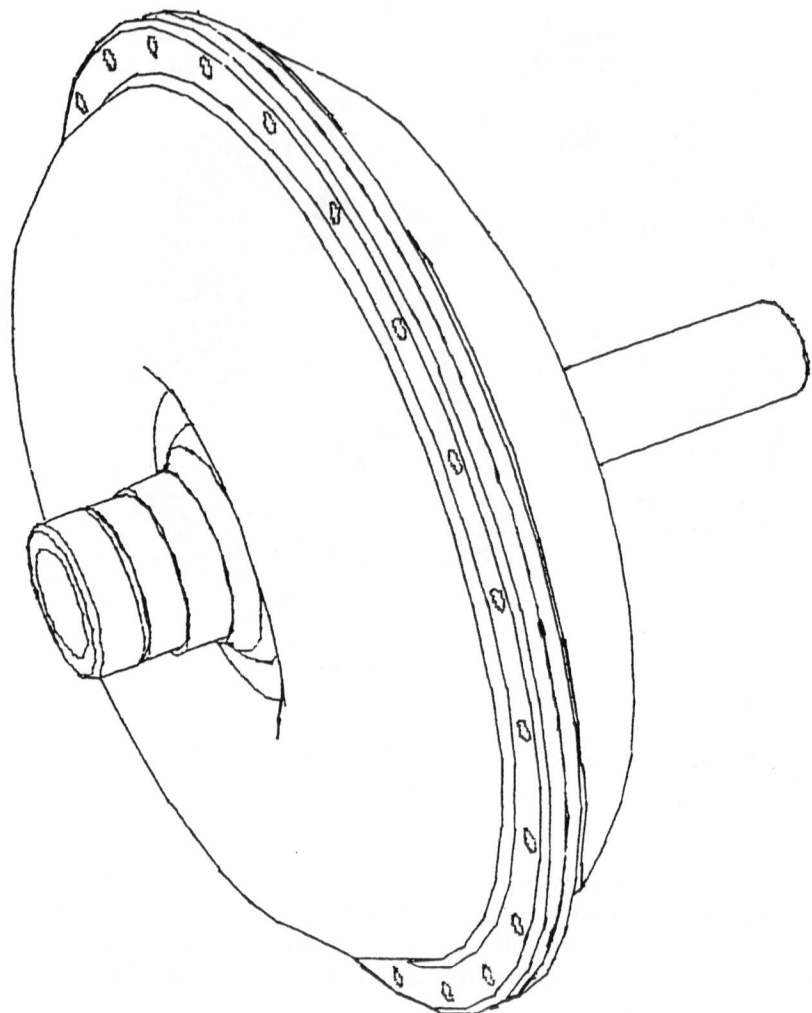

FIG. 3.32. General assembly.

effective method of cooling would be to recirculate the working fluid from within the coupling.

In this initial design, oil passages machined in the shaft were used to effect the circulation of the working fluid to external heat exchangers to cool it before re-entering the coupling. This set-up is shown in Fig. 3.34.

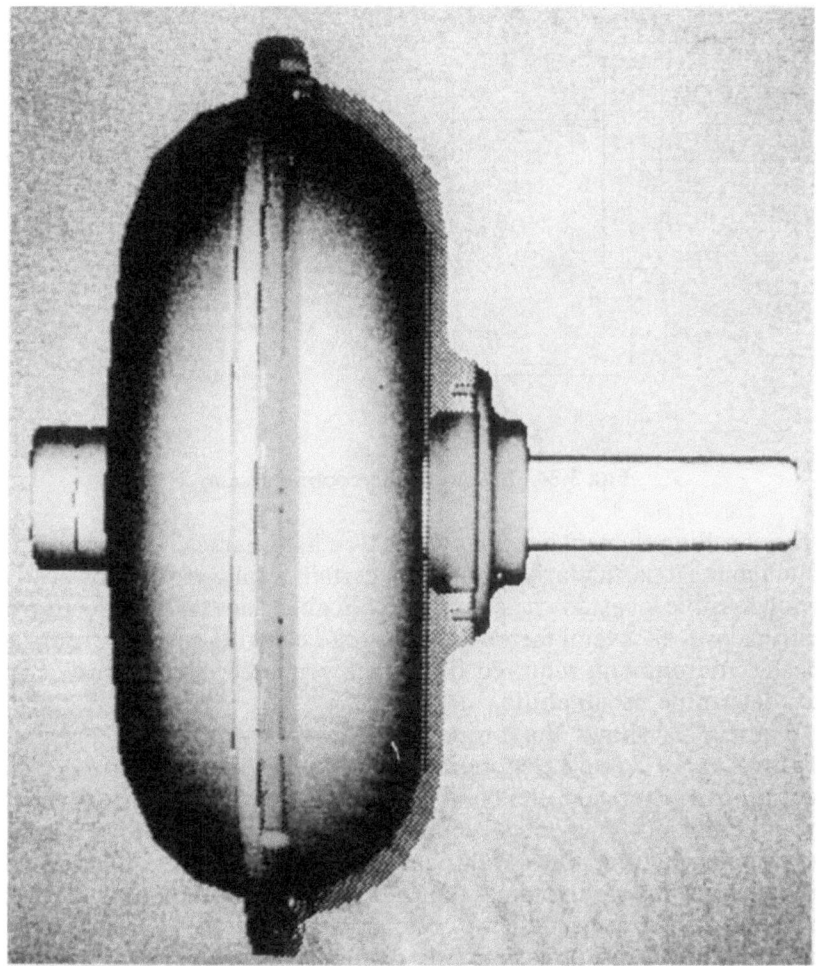

FIG. 3.33. General assembly with transparent housing.

Detailed calculations utilising fundamental heat transfer calculations, bearing calculations and seal selections are provided in Refs [3.18] to [3.20].

As a result of the rate of change of angular momentum of the fluid, axial forces will be introduced. These axial forces are reacted by the choice of a deep groove angular contact bearing [3.18]. The present design utilises rolling element type bearings. There are many different

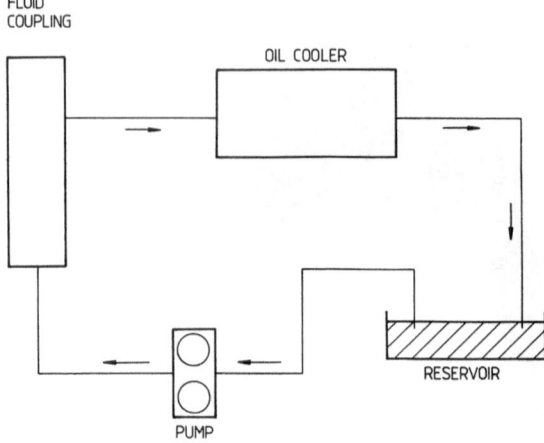

FIG. 3.34. Layout of the cooling system.

types of rolling element bearings. Each type has characteristic properties which make it particularly suitable for certain applications. However, it is not possible to lay down generally applicable rules for the selection of bearing type, as several factors, such as load carrying capacity, running speed, environment, required life, space provided, temperature, cost, etc., determine its suitability.

Figure 3.22 shows the proposed bearing layout. There are four bearings, A, B, C and D, supporting the fluid coupling. Bearing A is a deep groove spherical roller ball bearing which acts as a datum preventing any axial movement of the coupling. Bearing D is a cylindrical roller bearing which is capable of taking radial loads only, thus allowing one end of the coupling to float axially. If no axial movement is allowed for in the design, large loads will be experienced by the coupling due to the temperature rise. Bearing B is similar to D and also allows end float. Bearing C is an angular contact ball bearing which can carry high axial loads in one direction and only low axial loads in the opposite direction. The axial load imposed on bearing C is attributed to the axial force between the impeller and rotor resulting from the fluid movement. It is also capable of taking relatively low axial loads in the opposite direction thus fixing the position of the rotor relative to the impeller. The shaft and housing bearing surfaces are machined to the surface finish and limits of fit which are specified by the manufacturer, thus allowing each of the four bearings to be press fitted on to the shaft and into their respective housings. The two locating bearings (A and C) are additionally secured by retaining rings bolted to their respective casings.

Bearings B, C and D receive adequate cooling and lubrication from the working fluid of the coupling. Bearing A will be lubricated using a lithium based grease.

It is also proposed to use Shell Tellus Oil (T37) which possesses very low viscosity variation with temperature. It contains additives which produce strongly absorbed films on metal surfaces which protect them against corrosion. It also contains anti-wear elements and good oxidation stability. It is unlikely to pose any toxic hazard when used in this application, provided good standards of personal and industrial hygiene are observed.

3.10 EVALUATION OF COUPLING DESIGN

An evaluation of the above design reveals that it suffers from a number of drawbacks:

(i)   The present construction does not allow for relatively high reverse thrust during start-ups. This lack of restraint of the input shaft could result in touch-down between rotor and impeller leading to catastrophic consequences. It would therefore be necessary to positively constrain the shaft and ensure a specific gap (1–3 mm) between the elements.

(ii) The presence of oil passages in the shaft may affect its structural integrity and dynamic response of the assembly. The introduction of oilways will undoubtedly increase the stress levels in the shaft and consequently influence its fatigue performance.

(iii) The restricted dimensions of the oil passages in the shaft will restrict the flow rate of the working fluid.

(iv) The problems encountered in machining a dynamically balanced shaft with such longitudinal holes and the extra work needed to seal one end of these holes.

(v)   The introduction of a core in both impeller and rotor would reduce the cross-sectional area of each of the circulatory flow channels and thereby tend to reduce the mass flow rate $\dot{m}$ within the coupling and thus result in a reduction of the transmitted torque.

3.11 MODIFIED DESIGN

In order to overcome the above drawbacks, the following modified
design, which is shown in Fig. 3.35, is proposed. Detailed working drawings of the critical components, which include impeller, rotor, casing,
input and output shafts and oil inlet and outlet housings, are provided
in Fig. 3.36(a) through to Fig. 3.36(g). In this design, the oil circulation
is performed using two stationary housings which are directly connec-
ted to oil inlet and outlet passages of auxiliary coolers and coupling.
The working fluid which is pumped into the inlet housing is forced
through the oil passages into the rotor, then through the gap between the
elements and finally around the outside of the impeller to the outlet
housing. This arrangement ensures the continuous lubrication of the
bearings.

The solid model representation of the modified assembly is shown in
Fig. 3.37. It was felt that the design manufacture and maintenance of
both rotor and impeller could be improved by simply using semi-circular
disc vanes without a core as shown in Fig. 3.38.

The choice of the bearings was dictated by the need to accommodate
both static and dynamic loadings, to maintain a constant gap between
the rotor and impeller in order to avoid touch-down and to accommodate any movement due to thermal expansion. In the present design
study, angular contact deep groove and cylindrical roller bearings are
used; Fig. 3.39 shows the bearing arrangement adopted for the coupling.
Because of the sensitivity of the present exercise to misalignment, fits
and tolerances represent an important feature of the detailed working
drawings of the design.

3.12 THE NEED FOR ENGINEERING ANALYSIS

In order to evaluate and optimise the mechanical integrity of the previously mentioned design projects, some type of analysis is required.
Due to the complex nature of the geometrical features of the components
and/or the applied loads, the finite element method was mostly adopted; details of this are given in Chapters 4 and 5. In Chapter 4 a foundation for the finite element method is provided, while Chapter 5 is
concerned with the static structural stress analysis of the different components and with the vibration analysis of the different proposed
assemblies. In the development of the analysis techniques described, a

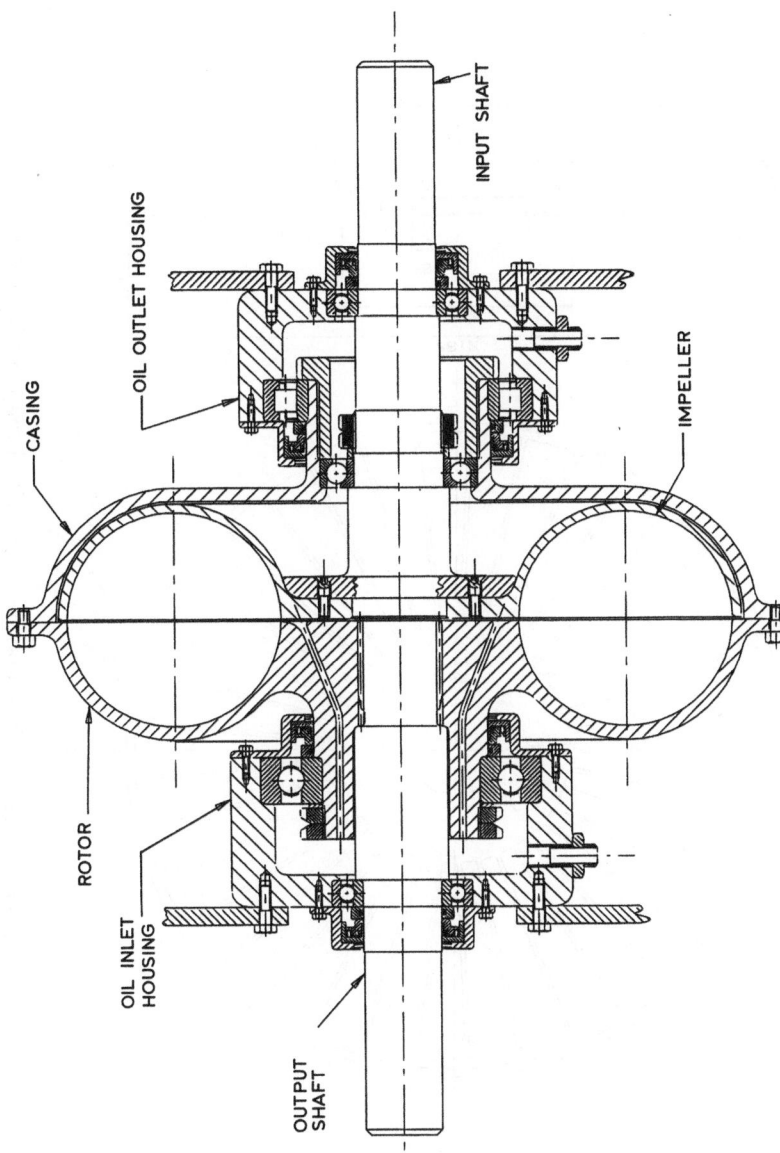

INPUT SHAFT

OIL OUTLET HOUSING

CASING

IMPELLER

ROTOR

OIL INLET
HOUSING

OUTPUT
SHAFT

FIG. 3.35. 2-D layout of the modified coupling general assembly using the McAuto Unigraphics system.

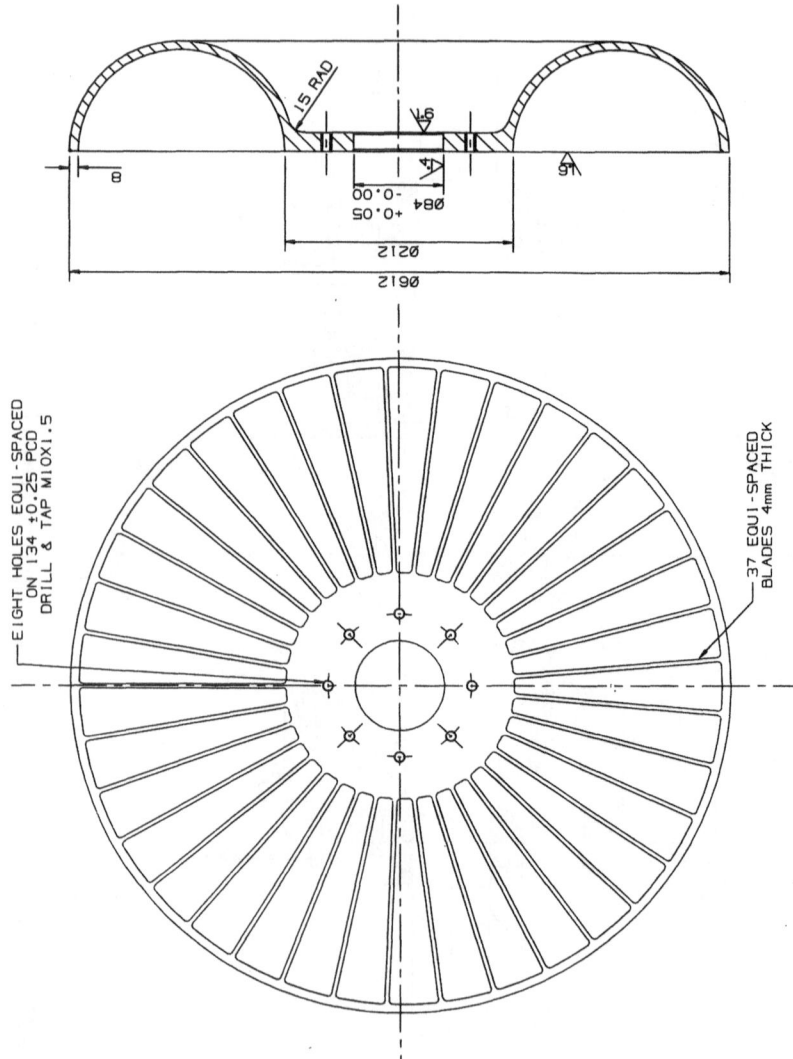

FIG. 3.36. Detailed working drawings of modified coupling: (a) impeller.

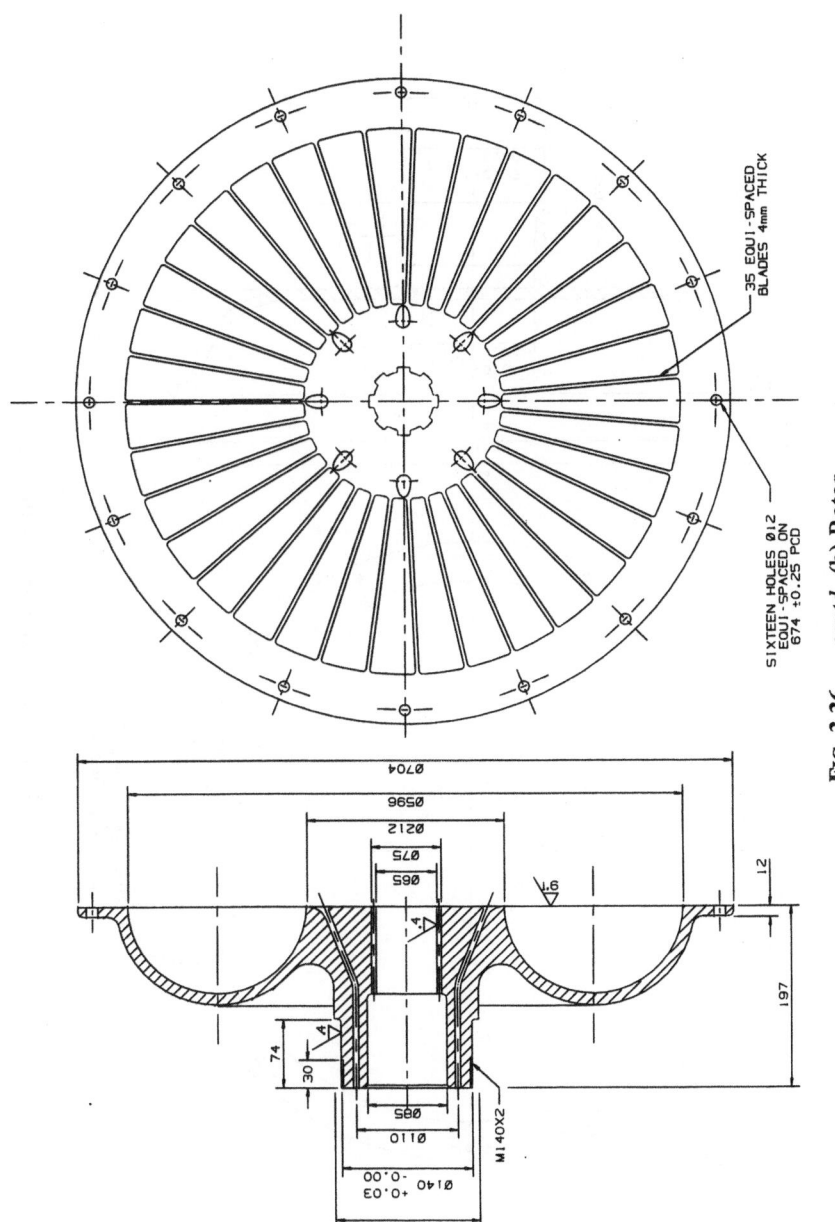

FIG. 3.36. — *contd.* (b) Rotor.

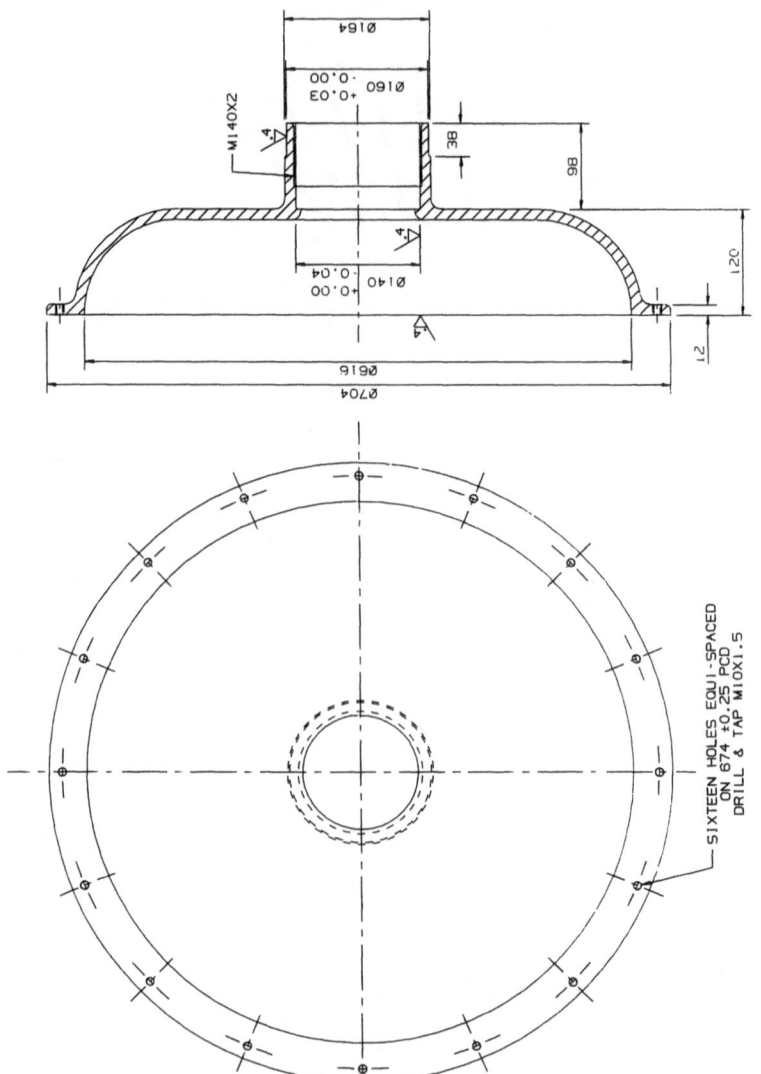

FIG. 3.36. — *contd.* (c) Casing.

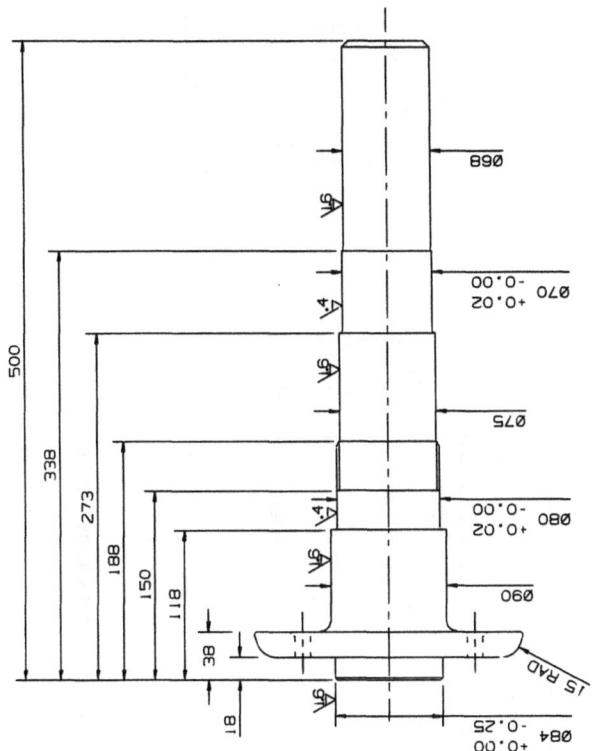

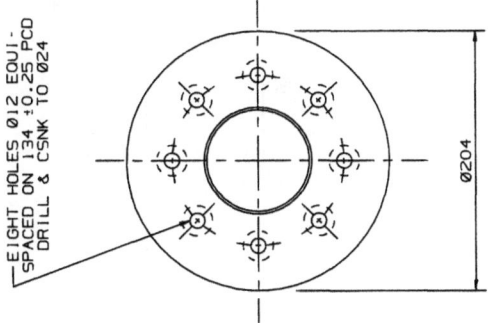

FIG. 3.36. — *contd.* (d) Input shaft.

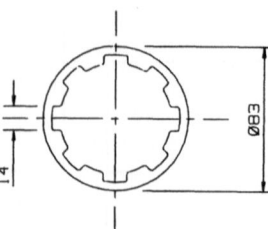

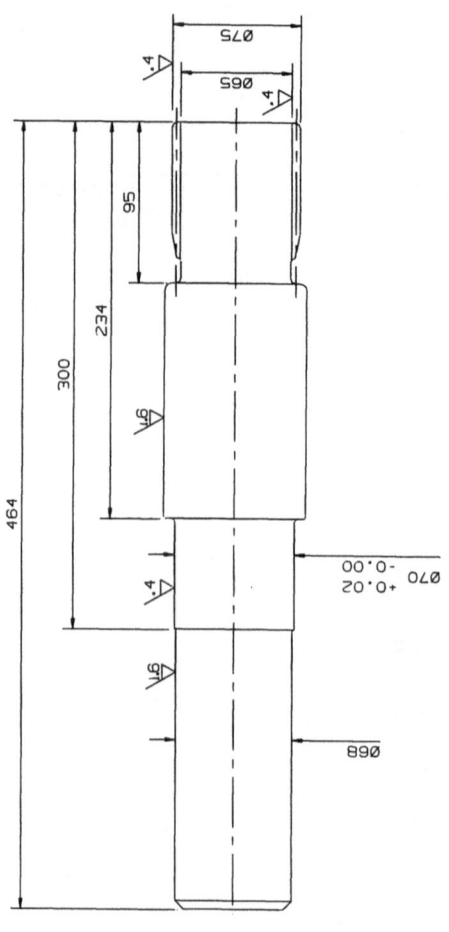

FIG. 3.36. — *contd.* (e) Output shaft.

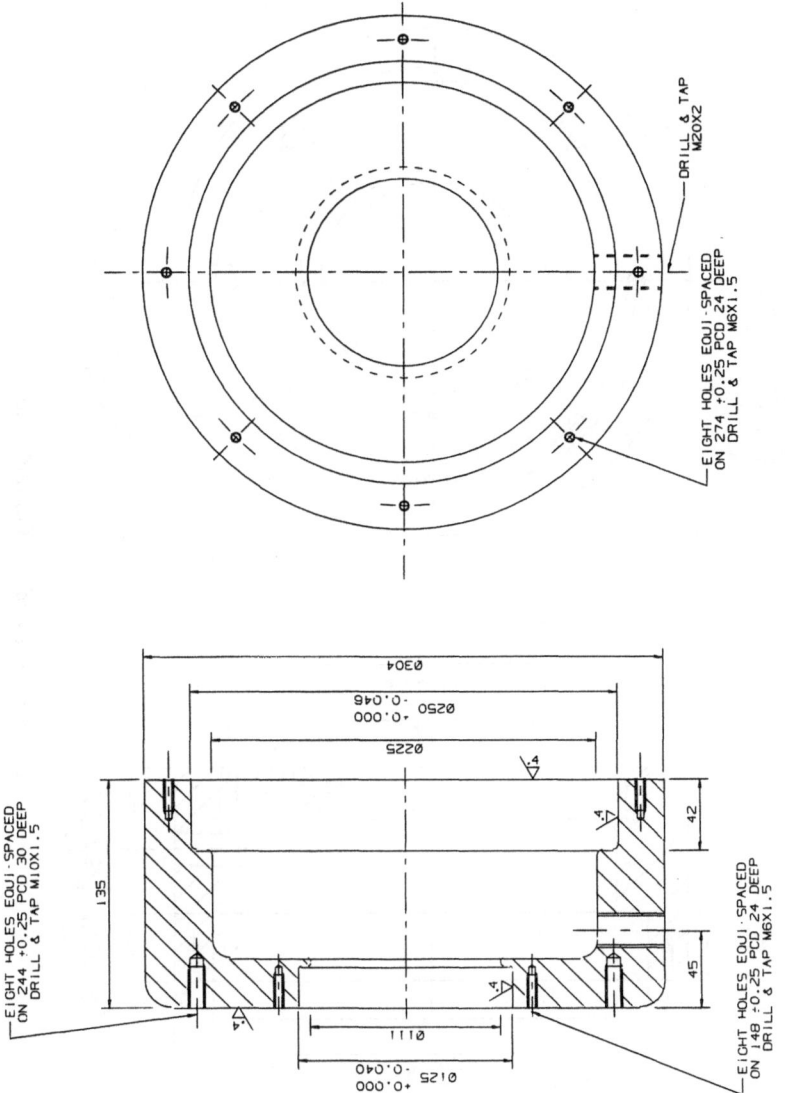

FIG. 3.36. — *contd.* (f) Oil inlet housing.

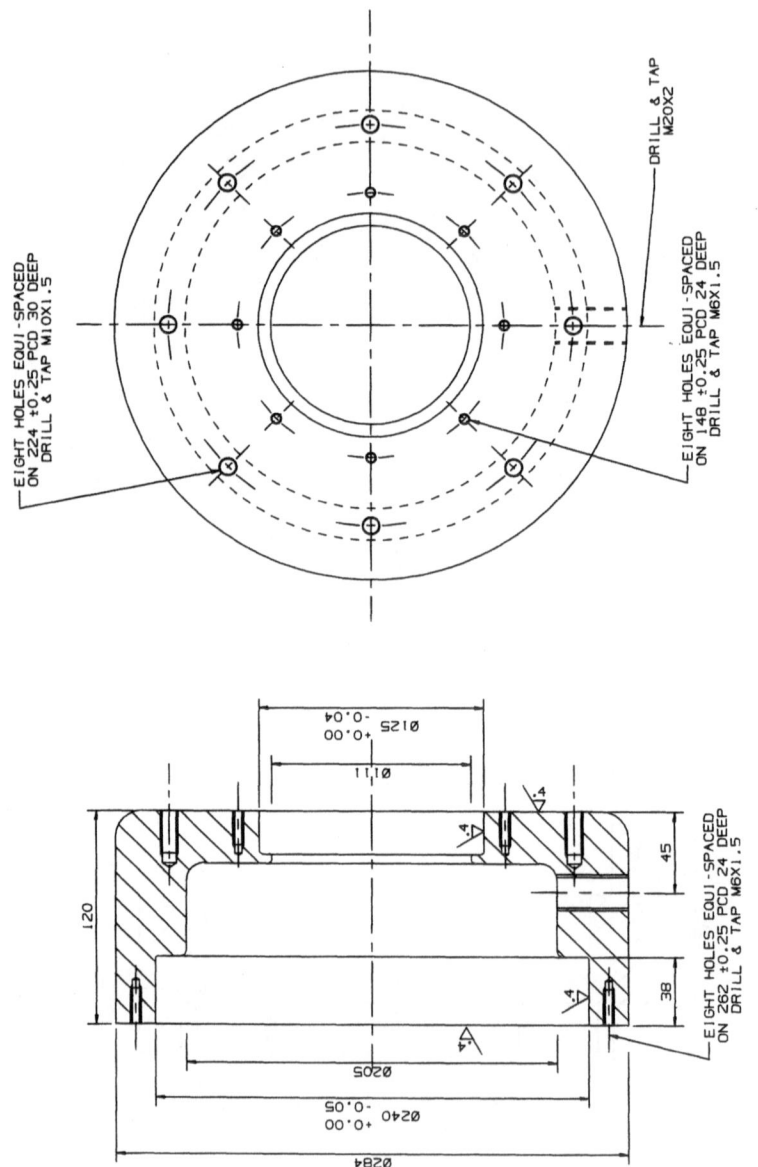

FIG. 3.36. — *contd.* (g) Oil outlet housing.

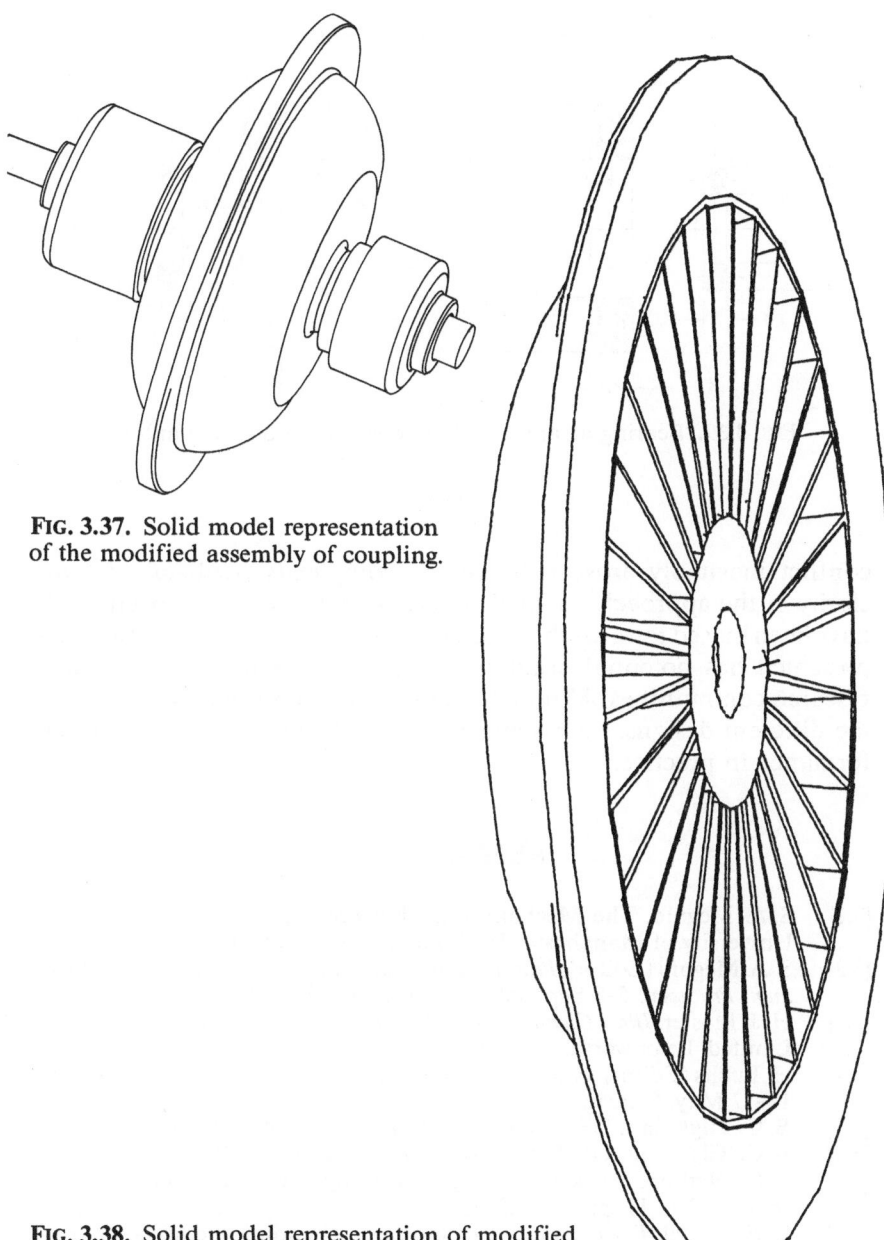

Fig. 3.37. Solid model representation
of the modified assembly of coupling.

Fig. 3.38. Solid model representation of modified
design of rotor (excluding bolt-holes).

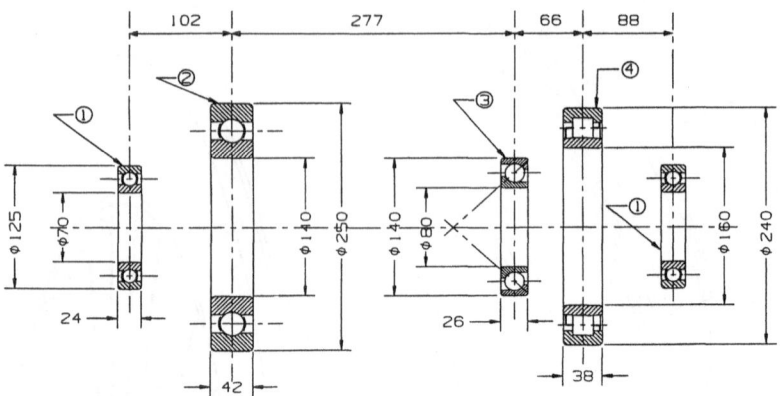

FIG. 3.39. Bearing arrangement adopted for modified coupling.

conflict inevitably arose between the complexity (realism) and the clarity of the approach adopted. Whatever sacrifices have been made have been biased towards the latter. However, it is believed that the work presented has potential usage in industrial as well as teaching and research communities. Although every attempt has been made to verify the different designs, no responsibility can be accepted for their performance in practice.

REFERENCES

[3.1] S. A. Meguid, The Mechanics of Shot-Peening Process, PhD Thesis, University of Manchester Institute of Science and Technology, 1975.

[3.2] S. A. Meguid (Editor), *Proc. First International Conference on Impact Treatment Processes,* 5–8 September, Cranfield, UK, 1983.

[3.3] H. J. Plaster, *Blast Cleaning and Allied Processes,* Industrial Newspapers Limited, Letchworth, UK, 1972.

[3.4] H. Fuchs (Editor), *Proc. Second International Conference on Shot-Peening,* 14–17 May, Chicago, 1984.

[3.5] B. C. Tilghman, US Patent No. 108,408, 18 October, 1870.

[3.6] B. C. Tilghman, UK Patent No. 2900, 3 November, 1870.

[3.7] E. G. Herbert, Work hardening of steel by abrasion, *J. Iron and Steel Institute,* No. 11, pp. 116–22, 1927.

[3.8] E. E. Weibel, The correlation of spring wire bending and torsion fatigue tests, *Trans-ASME,* **57,** pp. 28–36, 1935.

[3.9]   H. Berns and L. Weber, Crack initiation and growth in shot-peened and prestrained peened high strength steel, *Proc. Second International Conference on Shot-Peening,* 14–17 May, Chicago, pp. 84–9, 1984.

[3.10] R. B. Waterhouse, B. Noble and G. Leadbeater, The effect of shot-peening on the fretting-fatigue strength of an age-hardened aluminium alloy (2014A) and an austenitic stainless steel (En 58 A), *J. Mech. Working Technology,* **8**, pp. 147–53, 1983.

[3.11] M. Takemoto, Prevention of stress corrosion cracking of weldment by wet shot-peening, *Proc. Second International Conference on Shot-Peening,* 14–17 May, Chicago, pp. 39–42, 1984.

[3.12] W. Koehler, Influence of shot-peening with different peening materials on the stress corrosion and corrosion fatigue behaviour of a welded AlZn Mg-alloy, ibid., pp. 126–32, 1984.

[3.13] Tilghman Wheelabrator, UK Patent Nos. 937537 and 951257, 1975.

[3.14] Vacu-Blast International, UK Patent No. 1320641, 1972.

[3.15] *GEOMOD System Assembly Manual,* Structural Dynamics Research Corporation (SDRC), General Electric, CAE International, USA, 1983.

[3.16] F. J. Wallace, A. Whitfield and R. Sivalingam, A theoretical model for the performance of fully filled fluid couplings, *Int. J. Mechanical Sciences,* **20**, pp. 335–47, 1978.

[3.17] J. W. Qualman and E. L. Egbert, Fluid couplings — passenger car automatic transmission, *SAE Transmission Workshop Meeting, Passenger Car,* **5**, pp. 137–50, 1973.

[3.18] P. T. Coleman, The Computer-Aided Design of a Drive System for an Auxiliary Feed Pump, MSc Thesis, Cranfield Institute of Technology, 1984.

[3.19] W. May, Computer-Aided Analysis of a Fluid Coupling for an Auxiliary Feed Pump Application, MSc Thesis, Cranfield Institute of Technology, 1984.

[3.20] S. A. Meguid, W. May and P. T. Coleman, A computer-aided design study of a constant-fill hydraulic coupling, *Proc. International Conference on Computer-Aided Production Engineering,* 2–4 April, Edinburgh, pp. 157–60, 1986.

Chapter 4

Fundamentals of the Finite Element Method

4.1 INTRODUCTION

The finite element method is a powerful numerical tool for solving mathematical problems related to practical engineering situations. In the past, it was common practice to over-simplify such problems to the point where an analytical solution could be obtained. Because of the uncertainties associated with such a procedure, large safety factors were introduced in engineering design.

The finite element method has advanced from a numerical procedure for solving structural problems to a general numerical procedure for solving a differential equation or a system of differential equations. This advancement has been assisted by the development of high speed digital computers.

All finite element methods involve dividing the physical system into small sub-regions known as elements. Each element is essentially a simple unit, the behaviour of which can be readily analysed. The features of the overall system are accommodated for by using a large number of elements. Indeed, one of the attractions of the finite element method is the ease with which it can be applied to real engineering problems involving complex geometrical features. The price that must be paid for flexibility and simplicity of individual elements is in the amount of numerical computations required to solve the resulting sets of simultaneous algebraic equations.

Unfortunately, this simple, elegant and powerful method has so far been confined mainly to postgraduate research and teaching. This, together with the advanced mathematical skills demanded by most texts available at present, is responsible for creating a curious mystique about the subject which inhibits practising engineers from utilising what is after all a simple numerical method of analysis.

4.2 FUNDAMENTAL CONCEPT OF THE FINITE ELEMENT METHOD

The fundamental concept of the finite element method is that any continuous field variable, such as velocity, stress, pressure or temperature, can be approximated by a discrete model composed of a set of piecewise continuous field variables defined over a finite number of sub-domains, known as elements. These elements are interconnected at specified joints which are called nodes or nodal points. Since the actual variation of the field variable inside the continuum is not known, some approximating functions are needed to describe its variation. These approximating functions, which are also known as the interpolating functions, are defined in terms of the values of the field variable at the nodal points. When field equations, such as equilibrium or heat balance, for the whole body are written, the new unknowns will be the nodal values of the field variable. By solving the field equations, which are generally in the form of banded matrices, the nodal value of the field variable can be obtained throughout the assemblage of elements ([4.1] to [4.5]).

The general solution of an engineering problem can be detailed in a step-by-step procedure. This sequence of steps describes the actual solution process which is followed in setting up and solving equilibrium or heat balance problems. In this work, we concentrate on the approach adopted in structural mechanics problems, a summary of which is given below [4.6].

(i) Idealisation of structure: the geometrical features of the structure are simplified in order to accommodate sensible discretisation.

(ii) Discretisation of the structure: in this case, the body is subdivided into an equivalent system of finite elements. The type, size and number of elements is dictated by the geometrical features of the component, applied loads and restraints, accuracy needed and size of computer.

(iii) Choice of interpolation or displacement function: the assumed displacement function approximates the actual or exact distribution of the displacement field within the continuum. In general, the interpolation function is taken in the form of a polynomial, and practical considerations limit the number of terms that can be retained in the polynomial.

(iv) Derivation of element stiffness matrix: the stiffness matrix is composed of the coefficients of the equilibrium equations derived from the material and the geometric properties of an element and obtained by the use of the principle of minimum potential energy (equilibrium condition). The stiffness $[K^{(e)}]$ relates the displacements at the nodal points $\{u^{(e)}\}$ to the applied forces at the nodal points $\{F^{(e)}\}$, where (e) denotes the element number.

(v) Assembly of element equations for the overall discretised body: this process includes the assembly of the global stiffness matrix $[K]$ for the entire body from the individual element stiffness matrices $[K^{(e)}]$ and the global load vector $\{F\}$ from the element nodal force vectors $\{F^{(e)}\}$.

(vi) Solution for the unknown nodal displacements: the overall equilibrium equations have to be modified to account for the boundary conditions of the problem. After the incorporation of the boundary conditions, the global equilibrium equations can be expressed as

$$[K]\{u\} = \{F\} \tag{4.1}$$

For linear elastic problems, the displacement vector can be obtained easily. But for non-linear problems, the solution is obtained in a sequence of steps, each step involving the updating of the stiffness matrix $[K]$ and/or load vector $\{F\}$.

(vii) Computation of element strains and stresses from nodal displacements: having determined the primary unknowns (nodal displacements), it is often necessary to use these nodal displacements to determine the element strains and stresses by using the appropriate solid mechanics equations.

The application of the above-mentioned seven steps of the finite element analysis is best demonstrated by the following example, which is concerned with the determination of the displacement, strain and stress fields in the tapered bar of Fig. 4.1(a).

(i) Idealisation: the tapered bar is approximated by a limited number of one-dimensional bar elements as shown in Figs. 4.1(b) and (c).

(ii) Discretisation: the bar can be considered as an assemblage of three elements as shown in Fig. 4.1(d). By assuming the bar to be a one-dimensional structure, only axial displacement at any point in the element is considered. Since there are four nodes, the axial nodal displacements u_1, u_2, u_3 and u_4 will be taken as unknown.

(iii) Displacement model: in each of the elements, we assume a linear variation of axial displacement $u(x)$ so that

$$u(x) = a + bx$$

where a and b are constants. If we consider the respective end displacements $u_1^{(e)}$ and $u_2^{(e)}$ at $x = 0$ and $x = l^{(e)}$ as unknowns, we obtain

$$a = u_1^{(e)}, \text{ and } b = (u_2^{(e)} - u_1^{(e)})/l^{(e)}$$

where the superscript (e) denotes the element number. Thus

$$u_{(x)}^{(e)} = u_1^{(e)} + (u_2^{(e)} - u_1^{(e)})x/l^{(e)}$$

or

$$u_{(x)}^{(e)} = u_1^{(e)}N_1 + u_2^{(e)}N_2$$

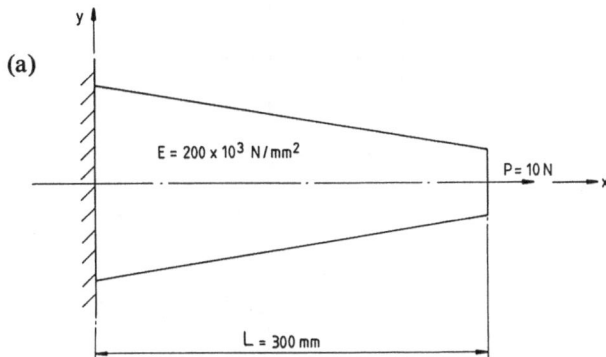

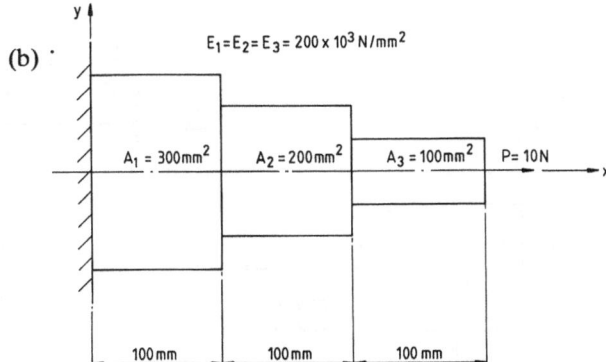

FIG. 4.1. Finite element analysis of a tapered bar: (a) definition of the problem; (b) idealisation of tapered bar.

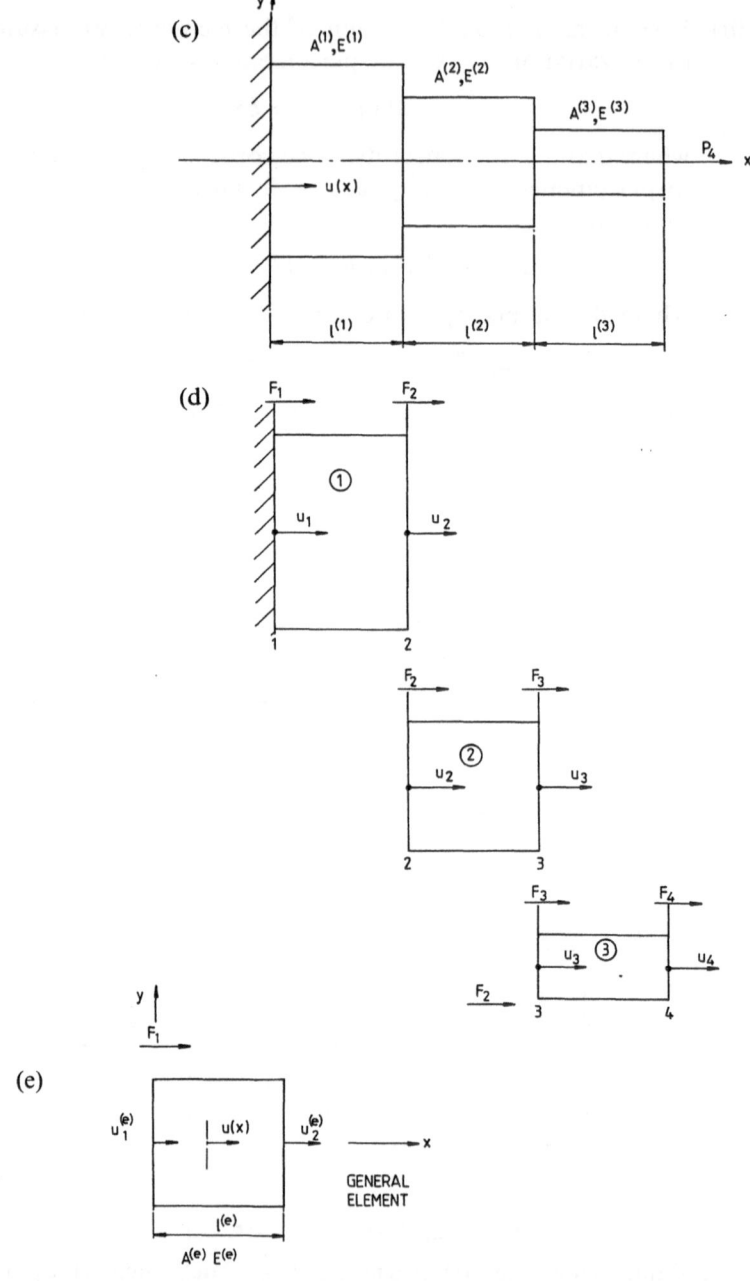

FIG. 4.1. — *contd.* (c) Discretisation of tapered bar; (d) assemblage of elements; (e) description of a general one-dimensional bar element.

where
$$N_1 = 1 - \frac{x}{l^{(e)}}$$

and
$$N_2 = \frac{x}{l^{(e)}}$$

and are known as shape functions.

(iv) Element stiffness matrix: the element stiffness matrices can be derived from the principle of minimum potential energy. Accordingly, we obtain the following expression for the potential energy of the bar (U) under axial deformation as

$$U = \text{strain energy} - \text{work done by external forces}$$

i.e.

$$U = \pi^{(1)} + \pi^{(2)} + \pi^{(3)} - W_p$$

where $\pi^{(e)}$ represents the strain energy of the (e)th element and W_p denotes the work done by the external forces. For the element shown in Fig. 4.1(e):

$$\pi^{(e)} = \int_0^{l^{(e)}} (1/2)\sigma^{(e)}\varepsilon^{(e)}dV = \frac{A^{(e)}E^{(e)}}{2} \int_0^{l^{(e)}} (\varepsilon^{(e)})^2 dx$$

where V = volume of (e)th element;
$A^{(e)}$ = cross-sectional area of the (e)th element;
$l^{(e)}$ = element length (= $L/3$);
$\sigma^{(e)}$ = stress in (e)th element;
$\varepsilon^{(e)}$ = strain in (e)th element;
$E^{(e)}$ = Young's Modulus of (e)th element.

Now, since
$$\varepsilon_{xx} = \frac{\partial u}{\partial x}, \text{ then}$$

$$\varepsilon^{(e)} = \frac{u_2^{(e)} - u_1^{(e)}}{l^{(e)}}$$

$$\therefore \quad \pi^{(e)} = \frac{A^{(e)}E^{(e)}}{2l^{(e)}} [(u_1^{(e)})^2 + (u_2^{(e)})^2 - 2u_1^{(e)}u_2^{(e)}]$$

This expression can be written in matrix form as

$$\pi^{(e)} = (1/2)\{u^{(e)}\}^T [K^{(e)}] \{u^{(e)}\}$$

where $\{u^{(e)}\}^T = \{u_1^{(e)} u_2^{(e)}\}$ which is the transpose of the nodal displacement vector of the (e)th element;

$$[K^{(e)}] = \frac{A^{(e)}E^{(e)}}{l^{(e)}} \begin{bmatrix} 1 & -1 \\ -1 & 1 \end{bmatrix}$$

which is called the element stiffness matrix; and

$$\{u^{(e)}\} = \begin{Bmatrix} u_1^{(e)} \\ u_2^{(e)} \end{Bmatrix}$$

which is the vector of the nodal displacements of the (e)th element:

$$\equiv \begin{Bmatrix} u_1 \\ u_2 \end{Bmatrix} \text{ for } e = 1,$$

$$\equiv \begin{Bmatrix} u_2 \\ u_3 \end{Bmatrix} \text{ for } e = 2, \text{ and}$$

$$\equiv \begin{Bmatrix} u_3 \\ u_4 \end{Bmatrix} \text{ for } e = 3.$$

The work done by external forces can be expressed as

$$W_p = u_1 F_1 + u_2 F_2 + u_3 F_3 + u_4 F_4$$

In the present example, F_1 = reaction at the fixed node, F_2 and $F_3 = 0$ and $F_4 = P = 10$ N.

If the entire bar is in equilibrium under the load vector

$$\{F\} = \begin{Bmatrix} F_1 \\ F_2 \\ F_3 \\ F_4 \end{Bmatrix}$$

then the principle of minimum potential energy gives

$$\frac{\partial U}{\partial u_i} = 0, i = 1, 2, 3 \text{ and } 4$$

This equation can be written as

$$\frac{\partial U}{\partial u_i} = \frac{\partial}{\partial u_i}\left(\sum_{e=1}^{3} \pi^{(e)} - W_p\right) = 0, i = 1, 2, 3 \text{ and } 4$$

i.e.

$$\sum_{e=1}^{3}\left([K^{(e)}]\{u^{(e)}\} - \{F^{(e)}\}\right) = 0$$

(v) Assemblage of element stiffness matrices:
The above equation can be written as

$$[K]\{u\} = \{F\}$$

where $[K]$ is called the global stiffness matrix $= \sum_{e=1}^{3} [K^{(e)}]$ and $\{u\}$ is known as the vector of global displacements. For the data given, the element matrices would be

$$[K^{(1)}] = \frac{A^{(1)}E^{(1)}}{l^{(1)}} \begin{bmatrix} 1 & -1 \\ -1 & 1 \end{bmatrix} = 10^5 \begin{array}{cc} u_1 & u_2 \\ \begin{bmatrix} 6 & -6 \\ -6 & 6 \end{bmatrix} \begin{array}{c} u_1 \\ u_2 \end{array} \end{array},$$

$$[K^{(2)}] = \frac{A^{(2)}E^{(2)}}{l^{(2)}} \begin{bmatrix} 1 & -1 \\ -1 & 1 \end{bmatrix} = 10^5 \begin{array}{cc} u_2 & u_3 \\ \begin{bmatrix} 4 & -4 \\ -4 & 4 \end{bmatrix} \begin{array}{c} u_2 \\ u_3 \end{array} \end{array},$$

$$\text{and } [K^{(3)}] = \frac{A^{(3)}E^{(3)}}{l^{(3)}} \begin{bmatrix} 1 & -1 \\ -1 & 1 \end{bmatrix} = 10^5 \begin{array}{cc} u_3 & u_4 \\ \begin{bmatrix} 2 & -2 \\ -2 & 2 \end{bmatrix} \begin{array}{c} u_3 \\ u_3 \end{array} \end{array}.$$

Since the displacements of the left and right nodes of the first element are u_1 and u_2, the rows and columns of the stiffness matrix corresponding to these unknowns are identified as shown above. Similarly, the rows and columns of the stiffness matrices of the other elements are identified with the unknown displacements u_2, u_3 and u_4.

The overall or global stiffness of the bar can be obtained by assembling the three element stiffness matrices. Since there are four nodal unknown displacements (u_1, u_2, u_3 and u_4) the global stiffness matrix $[K]$ will be of order four. The elements of the global stiffness matrix corresponding to the unknown displacements u_1, u_2, u_3 and u_4, are obtained by adding the corresponding terms as shown below:

$$[K] = 10^5 \begin{array}{cccc} u_1 & u_2 & u_3 & u_4 \\ \begin{bmatrix} 6 & -6 & 0 & 0 \\ -6 & 6+4 & -4 & 0 \\ 0 & -4 & 4+2 & -2 \\ 0 & 0 & -2 & 2 \end{bmatrix} & \begin{array}{c} u_1 \\ u_2 \\ u_3 \\ u_4 \end{array} \end{array}$$

$$= 10^5 \begin{bmatrix} 6 & -6 & 0 & 0 \\ -6 & 10 & -4 & 0 \\ 0 & -4 & 6 & -2 \\ 0 & 0 & -2 & 2 \end{bmatrix}$$

The global load vector can be written as

$$\{F\} = \begin{Bmatrix} F_1 \\ F_2 \\ F_3 \\ F_4 \end{Bmatrix} = \begin{Bmatrix} 10 \\ 0 \\ 0 \\ 10 \end{Bmatrix}$$

where F_1 denotes the reaction at node 1. Thus the overall equilibrium equations become

$$10^5 \begin{bmatrix} 6 & -6 & 0 & 0 \\ -6 & 10 & -4 & 0 \\ 0 & -4 & 6 & -2 \\ 0 & 0 & -2 & 2 \end{bmatrix} \begin{Bmatrix} u_1 \\ u_2 \\ u_3 \\ u_4 \end{Bmatrix} = \begin{Bmatrix} F_1 \\ 0 \\ 0 \\ 10 \end{Bmatrix}$$

(vi) Solution for displacements:

In order to avoid a singular global stiffness matrix, we incorporate the known boundary condition; $u_1 = 0$. The reduced equilibrium equations can be written as

$$10^5 \begin{bmatrix} 10 & -4 & 0 \\ -4 & 6 & -2 \\ 0 & -2 & 2 \end{bmatrix} \begin{Bmatrix} u_2 \\ u_3 \\ u_4 \end{Bmatrix} = \begin{Bmatrix} 0 \\ 0 \\ 10 \end{Bmatrix}.$$

The solution of the above linear equations gives

$$u_2 = 1{\cdot}66 \times 10^{-5} \text{ mm}$$
$$u_3 = 4{\cdot}166 \times 10^{-5} \text{ mm}$$
$$u_4 = 9{\cdot}166 \times 10^{-5} \text{ mm}$$

(vii) Element strains and stresses:

From the known displacements, the strains in the elements can be computed as

$$\varepsilon_{xx}{}^{(1)} = \frac{\partial u}{\partial x} \text{ for element (1)} = \frac{u_2{}^{(1)} - u_1{}^{(1)}}{l^{(1)}} = 1{\cdot}66 \times 10^{-7}$$

$$\varepsilon_{xx}^{(2)} = \frac{\partial u}{\partial x} \text{ for element (2)} = \frac{u_3^{(2)} - u_2^{(2)}}{l^{(2)}} = 3 \times 10^{-7}$$

$$\varepsilon_{xx}^{(3)} = \frac{\partial u}{\partial x} \text{ for element (3)} = \frac{u_4^{(3)} - u_3^{(3)}}{l^{(3)}} = 5 \times 10^{-7}$$

The stresses in the elements are given by

$$\sigma_{xx}^{(1)} = E^{(1)}\varepsilon_{xx}^{(1)} = 332 \text{ N/m}^2$$

$$\sigma_{xx}^{(2)} = E^{(2)}\varepsilon_{xx}^{(2)} = 600 \text{ N/m}^2$$

$$\sigma_{xx}^{(3)} = E^{(3)}\varepsilon_{xx}^{(3)} = 1000 \text{ N/m}^2$$

4.3 FUNDAMENTALS OF LINEAR ELASTICITY

The derivation of the stiffness matrix of the bar elements discussed in the above example were deliberately restricted within the confines of elementary structural mechanics with which the beginner is familiar. In the following section, we shall present a general procedure, based upon the principle of virtual work, from which the stiffness and other characteristics of any finite element can be derived. As the under-standing and application of this procedure is facilitated by some knowledge of elasticity, we shall discuss briefly some of the essential aspects of linear elasticity which are of interest in the general context of finite elements. For simplicity, we devote our attention to cartesian co-ordinates.

4.3.1 Stresses and strains
Throughout this text the symbols σ and τ are used to denote stress. Individual components of stress are indicated by double subscripts, as follows:

Direct stresses $\quad \sigma_{xx}, \sigma_{yy}, \sigma_{zz}$

Shear stresses $\quad \tau_{xy}, \tau_{yz}, \tau_{zx}, \tau_{yx}, \tau_{zy}, \tau_{xz}$

The first subscript defines the direction of the outward normal to the surface on which it acts and the second denotes the direction of the stress, as shown in Fig. 4.2.

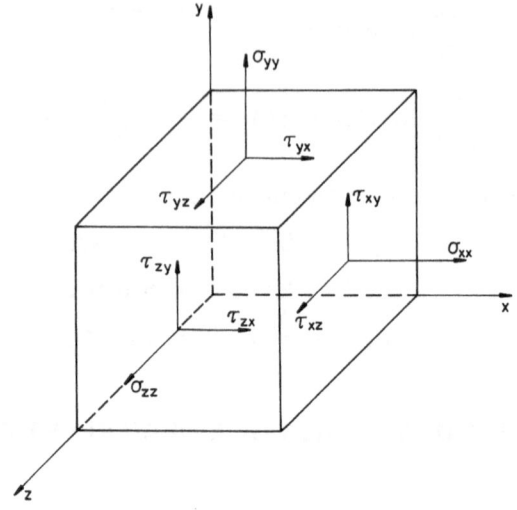

FIG. 4.2. Stress components referred to cartesian co-ordinates.

For rotational equilibrium to be maintained, the shear stresses must be complementary, i.e.

$$\tau_{xy} = \tau_{yx}, \tau_{yz} = \tau_{zy}, \tau_{zx} = \tau_{xz} \qquad (4.2)$$

The components of the displacement for a solid are denoted by u, v and w in the co-ordinate directions x, y and z, respectively. Using the same double subscript notation, the direct components of strain can be determined as

$$\varepsilon_{xx} = \frac{\partial u}{\partial x}, \varepsilon_{yy} = \frac{\partial v}{\partial y}, \varepsilon_{zz} = \frac{\partial w}{\partial z} \qquad (4.3)$$

and the engineering shear strains as

$$\gamma_{xy} = \gamma_{yx} = \frac{\partial u}{\partial y} + \frac{\partial v}{\partial x} \qquad (4.4)$$

$$\gamma_{yz} = \gamma_{zy} = \frac{\partial v}{\partial z} + \frac{\partial w}{\partial y} \qquad (4.5)$$

$$\gamma_{zx} = \gamma_{xz} = \frac{\partial u}{\partial z} + \frac{\partial w}{\partial x} \qquad (4.6)$$

4.3.2 Stress–strain relationships

Unless otherwise stated, we shall assume that the material of the body is linearly elastic, isotropic and homogeneous, so that its elastic properties are completely identified by the mutually independent constants E and v, denoting Young's modulus and Poisson's ratio, respectively. In this case, the normal strains are related to the stresses via Hooke's law as follows:

$$\varepsilon_{xx} = \frac{1}{E}[\sigma_{xx} - v(\sigma_{yy} + \sigma_{zz})] \tag{4.7}$$

$$\varepsilon_{yy} = \frac{1}{E}[\sigma_{yy} - v(\sigma_{xx} + \sigma_{zz})] \tag{4.8}$$

and

$$\varepsilon_{zz} = \frac{1}{E}[\sigma_{zz} - v(\sigma_{xx} + \sigma_{yy})] \tag{4.9}$$

Unlike the normal strains, the shear strains are mutually independent, and related to their respective stresses through the following relations:

$$\gamma_{xy} = \tau_{xy}/G \tag{4.10}$$

$$\gamma_{yz} = \tau_{yz}/G \tag{4.11}$$

$$\gamma_{zx} = \tau_{zx}/G \tag{4.12}$$

where G is the modulus of rigidity of the component and is given by

$$G = E/2(1 + v) \tag{4.13}$$

It is convenient to write the above six strain components in a matrix form, as follows:

$$
\begin{Bmatrix} \varepsilon_{xx} \\ \varepsilon_{yy} \\ \varepsilon_{zz} \\ \gamma_{xy} \\ \gamma_{yz} \\ \gamma_{zx} \end{Bmatrix} = \frac{1}{E}
\begin{bmatrix}
1 & -v & -v & 0 & 0 & 0 \\
-v & 1 & -v & 0 & 0 & 0 \\
-v & -v & 1 & 0 & 0 & 0 \\
0 & 0 & 0 & 2(1+v) & 0 & 0 \\
0 & 0 & 0 & 0 & 2(1+v) & 0 \\
0 & 0 & 0 & 0 & 0 & 2(1+v)
\end{bmatrix}
\begin{Bmatrix} \sigma_{xx} \\ \sigma_{yy} \\ \sigma_{zz} \\ \tau_{xy} \\ \tau_{yz} \\ \tau_{zx} \end{Bmatrix}
\tag{4.14}
$$

By inversion, we can see the above equations in the following alternative form:

$$
\begin{Bmatrix} \sigma_{xx} \\ \sigma_{yy} \\ \sigma_{zz} \\ \tau_{xy} \\ \tau_{yz} \\ \tau_{zx} \end{Bmatrix} = \begin{bmatrix} \lambda + 2G & \lambda & \lambda & 0 & 0 & 0 \\ \lambda & \lambda + 2G & \lambda & 0 & 0 & 0 \\ \lambda & \lambda & \lambda + 2G & 0 & 0 & 0 \\ 0 & 0 & 0 & G & 0 & 0 \\ 0 & 0 & 0 & 0 & G & 0 \\ 0 & 0 & 0 & 0 & 0 & G \end{bmatrix} \begin{Bmatrix} \varepsilon_{xx} \\ \varepsilon_{yy} \\ \varepsilon_{zz} \\ \gamma_{xy} \\ \gamma_{yz} \\ \gamma_{zx} \end{Bmatrix}
$$

$$(4.15)$$

in which λ is the well-known lamé coefficient, and is given by

$$\lambda = \frac{vE}{(1+v)(1-2v)} \tag{4.16}$$

Equations (4.15) state Hooke's law in three dimensions which can be written in the following compact form:

$$\{\sigma\} = [D]\{\varepsilon\} \tag{4.17}$$

in which $[D]$ is the elasticity matrix represented by the three-dimensional array of eqns (4.15),

$$\{\varepsilon\} = \{\varepsilon_{xx} \; \varepsilon_{yy} \; \varepsilon_{zz} \; \gamma_{xy} \; \gamma_{yz} \; \gamma_{zx}\}^{\mathrm{T}} \tag{4.18}$$

and $\{\sigma\} = \{\sigma_{xx} \; \sigma_{yy} \; \sigma_{zz} \; \tau_{xy} \; \tau_{yz} \; \tau_{zx}\}^{\mathrm{T}}$

4.3.3 Plane stress and plane strain
If a body consists of two parallel planes a constant thickness apart and bounded by any closed surface, it is said to be a plane body. Associated with this type of body there is a particular class of problems within the general theory of elasticity which are termed plane elastic problems, and these allow a number of simplifying assumptions in their treatment [4.7] and [4.8].

Plane stress problems
A plane stress condition can be used to analyse elasticity problems in which the body is subjected to in-plane forces and one of its dimensions is very small compared to the rest. A typical problem is shown in Fig. 4.3 in which

$$\sigma_{zz} = \tau_{xz} = \tau_{yz} = 0 \tag{4.19}$$

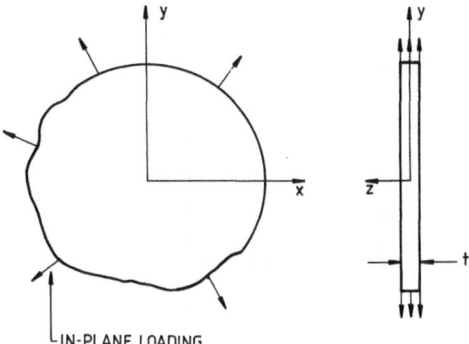

FIG. 4.3. Plane stress configuration.

Furthermore, since t is sufficiently small compared with the other dimensions of the body, and provided that the forces applied at the edges are uniform across t, then we may assume that the above condition is also valid within the cross-section of the body. In this case, the remaining stresses σ_{xx}, σ_{yy} and τ_{xy} are assumed not to vary across the thickness t. The problem thus becomes a two-dimensional one in which the z-axis can be ignored, and the stress–strain relations reduce to

$$\varepsilon_{xx} = \frac{1}{E}(\sigma_{xx} - v\sigma_{yy}) \tag{4.20}$$

$$\varepsilon_{yy} = \frac{1}{E}(\sigma_{yy} - v\sigma_{xx}) \tag{4.21}$$

and $$\gamma_{xy} = \tau_{xy}/G \tag{4.22}$$

and a state of plane stress is said to exist. In this case, the elasticity matrix $[D]$ is reduced to

$$[D] = \frac{E}{1 - v^2} \begin{bmatrix} 1 & v & 0 \\ v & 1 & 0 \\ 0 & 0 & \dfrac{1-v}{2} \end{bmatrix} \tag{4.23}$$

Plane strain problems
Figure 4.4 shows a solid body whose cross-section is uniform in the z-direction. Provided its length in this direction is large, the section OACB can be regarded as being remote from the ends.

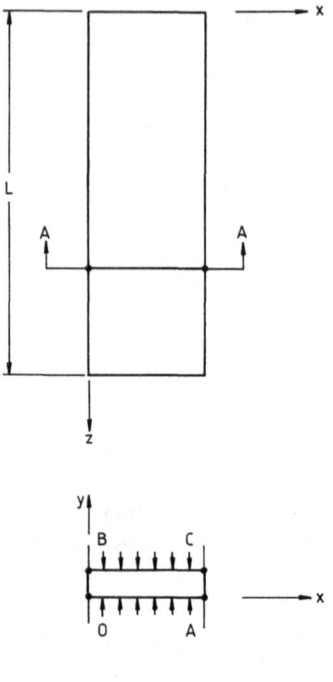

SECTION A-A

FIG. 4.4. Plane strain configuration.

Assuming the applied loads are in the *xy*-plane, the resulting state of strain at such a section is two-dimensional, with $w = 0$ and $\dfrac{\partial}{\partial z} = 0$.

As a result:

$$\varepsilon_{zz} = \gamma_{xz} = \gamma_{yz} = 0 \qquad (4.24)$$

and a state of plane strain is said to exist. In this case, the elasticity matrix $[D]$ is reduced to

$$\begin{bmatrix} \lambda + 2G & \lambda & 0 \\ \lambda & \lambda + 2G & 0 \\ 0 & 0 & G \end{bmatrix} \qquad (4.25)$$

4.3.4 Equilibrium equations

The differential equations of equilibrium for the three co-ordinate directions can be expressed as

$$\frac{\partial \sigma_{xx}}{\partial x} + \frac{\partial \tau_{xy}}{\partial y} + \frac{\partial \tau_{xz}}{\partial z} + \bar{X} = 0 \qquad (4.26)$$

$$\frac{\partial \sigma_{yy}}{\partial y} + \frac{\partial \tau_{xy}}{\partial x} + \frac{\partial \tau_{yz}}{\partial z} + \bar{Y} = 0 \qquad (4.27)$$

$$\frac{\partial \sigma_{zz}}{\partial z} + \frac{\partial \tau_{xz}}{\partial x} + \frac{\partial \tau_{yz}}{\partial y} + \bar{Z} = 0 \qquad (4.28)$$

where $\bar{X}$, $\bar{Y}$ and $\bar{Z}$ are the local components of the body forces per unit volume acting on the continuum in the co-ordinate directions. Both inertia and gravity forces are the most common causes of such forces.

4.3.5 Compatibility equations

Strains must be compatible with each other. This effectively means that no discontinuities such as holes or overlap of material exist in the body. The mathematical formulation of this condition is as follows:

$$\frac{\partial^2 \varepsilon_{xx}}{\partial y^2} + \frac{\partial^2 \varepsilon_{yy}}{\partial x^2} = 2 \frac{\partial^2 \varepsilon_{xy}}{\partial x\, \partial y} \qquad (4.29)$$

$$\frac{\partial^2 \varepsilon_{yy}}{\partial z^2} + \frac{\partial^2 \varepsilon_{zz}}{\partial y^2} = 2 \frac{\partial^2 \varepsilon_{yz}}{\partial y\, \partial z} \qquad (4.30)$$

$$\frac{\partial^2 \varepsilon_{zz}}{\partial x^2} + \frac{\partial^2 \varepsilon_{xx}}{\partial z^2} = 2 \frac{\partial^2 \varepsilon_{xz}}{\partial x\, \partial z} \qquad (4.31)$$

$$\frac{\partial}{\partial x}\left(-\frac{\partial \varepsilon_{yz}}{\partial x} + \frac{\partial \varepsilon_{zx}}{\partial y} + \frac{\partial \varepsilon_{xy}}{\partial z} \right) = \frac{\partial^2 \varepsilon_{xx}}{\partial y\, \partial z} \qquad (4.32)$$

$$\frac{\partial}{\partial y}\left(-\frac{\partial \varepsilon_{zx}}{\partial y} + \frac{\partial \varepsilon_{xy}}{\partial z} + \frac{\partial \varepsilon_{yz}}{\partial x} \right) = \frac{\partial^2 \varepsilon_{yy}}{\partial z\, \partial x} \qquad (4.33)$$

$$\frac{\partial}{\partial z}\left(-\frac{\partial \varepsilon_{xy}}{\partial z} + \frac{\partial \varepsilon_{yz}}{\partial x} + \frac{\partial \varepsilon_{zx}}{\partial y} \right) = \frac{\partial^2 \varepsilon_{zz}}{\partial x\, \partial y} \qquad (4.34)$$

where ε_{xy}, ε_{yz} and ε_{zx} are the tensor components of the shear strains.

4.4 PRINCIPLES OF FINITE ELEMENTS

4.4.1 Discretisation of a body

The discretisation of a body into sub-regions is the first of a series of steps that must be performed when solving an engineering problem. This particular step does not have a theoretical basis. It is an art and depends on the use of engineering judgement. The application of poor or improper judgement will produce inaccurate results even though all the other steps are rigorously adhered to [4.2].

The discretisation of a body involves the decision as to the number, size and shape of elements used to model the body. The general objective of such a discretisation is to divide the body into elements sufficiently small so that the simple displacement functions can adequately approximate the solution. At the same time, one must remember that too fine a mesh will lead to extra computational effort. The judicious sub-division of a domain comes with experience. However, several general rules can be used to guide the inexperienced. These rules and some general advice relative to the physical process of discretisation are discussed in the present section. However, let us begin by providing some of the common elements used in the discretisation of engineering components.

4.4.2 Types of elements

One-dimensional elements
When the geometry, material properties and such dependent variables as temperature, displacement or stress can be described in terms of only one independent spatial co-ordinate, a one-dimensional element can be used. The simplest one-dimensional element has two nodes, one at each end. Although this element has a cross-sectional area, it is generally shown schematically as a line segment, as shown in Fig. 4.5(a). The most common higher order elements are the three-noded (quadratic) and four-noded (cubic) elements shown in Fig. 4.5(b) and (c). The most common usages of this element are in the one-dimensional heat transfer problems and truss and frame analysis.

Two-dimensional elements
In Section 4.3.3 we demonstrated that some problems in solid mechanics can be approximated by a two-dimensional formulation. Among these problems are plane stress, plane strain and plate bending. There are two

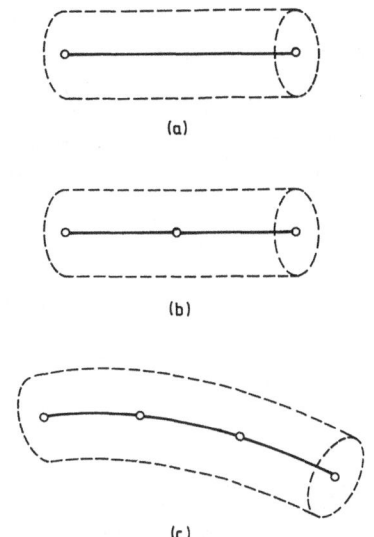

FIG. 4.5. One-dimensional finite elements.

general element families used to model the two-dimensional domain, the triangle and the quadrilateral of Fig. 4.6(a) and (b). In this case, the linear elements in each family have straight sides, but the higher order elements, quadratic and cubic, can have either straight or curved sides or both. The ability to model curved boundaries is made possible by the addition of midside nodes. Both families of elements can be used within the same body provided that both have the same number of nodes along a side in order to ensure compatibility between the nodes.

Three-dimensional solid elements

The most common three-dimensional elements are variations of the two-dimensional elements, namely the tetrahedron and the hexahedron. A tetrahedron has four primary external nodes, while a general hexahedron and a rectangular prism have eight primary external nodes as depicted in Fig. 4.7. In this case, the linear elements are restricted to plane surfaces, while the higher order elements can have curved surfaces for their sides.

Some problems which are actually three-dimensional can be approximated by two independent co-ordinates. Such problems can be approximated by using the axisymmetric element shown in Fig. 4.8. The

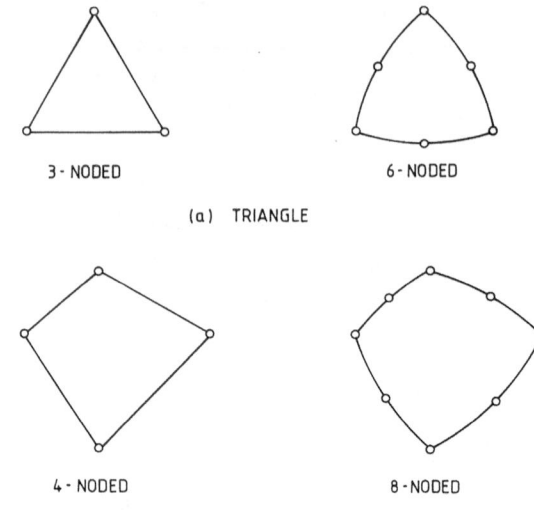

3 - NODED   6 - NODED

(a)   TRIANGLE

4 - NODED   8 - NODED

(b)   QUADRILATERAL

FIG. 4.6. Some two-dimensional finite elements.

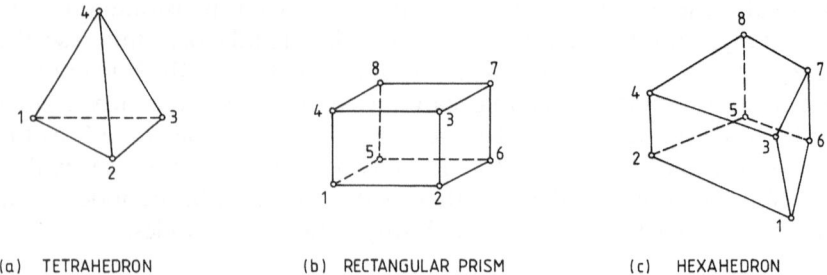

(a)   TETRAHEDRON   (b)   RECTANGULAR PRISM   (c)   HEXAHEDRON

FIG. 4.7. Some three-dimensional finite elements.

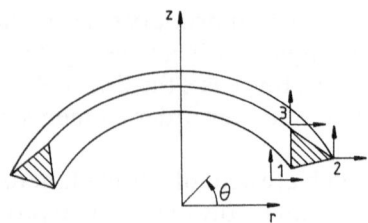

FIG. 4.8. Axisymmetric toroidal element.

problems that possess axial-symmetry, like pistons, pressure vessels, nozzles, rotating discs, fall into this category. It is convenient to express these problems in terms of the cylindrical co-ordinate system. Because of symmetry, the stress components are independent of the tangential direction (θ). In this case, the elasticity matrix $[D]$ is given by

$$[D] = \frac{E}{(1+v)(1-2v)} \begin{bmatrix} 1-v & v & v & 0 \\ v & 1-v & v & 0 \\ v & v & 1-v & 0 \\ 0 & 0 & 0 & \dfrac{1-2v}{2} \end{bmatrix} \quad (4.35)$$

4.4.3 Division of the body into elements

The actual discretisation process can be divided into two parts: the first deals with the division of the body into elements and the second deals with the labelling of the elements and the numbering of the nodes; the latter sounds quite simple but is complicated by our desire to increase the computational efficiency [4.1], [4.2] and [4.6].

Often the type of elements to be used will be evident from the physical characteristics of the problem itself. For example, one-dimensional elements are used if the problem involves the analysis of a truss structure or one-dimensional heat transfer problems or the dynamic response of some rotating machines. Similarly, in the case of stress analysis of engineering components, the modelling of the component can be done using three-dimensional solid elements as shown in Fig. 4.9. In some cases, however, the type of element to be used for the division of the body may not be apparent. In this situation, the type of element is dictated by the number of degrees of freedom, accuracy required, computational power available and other related sensible parameters. In certain problems, the division of the body cannot be made using only one type of element, e.g. bending of a plate which is stiffened by edge beams as in aircraft wings. Here, two-dimensional elements supplemented by beam elements will be used [4.5].

The size of the element should be selected with care. If the size of the element is relatively small, then the final solution is expected to be more accurate. However, this is always associated with increased computational time and hence increased cost. In true engineering situations, most bodies have zones in which pronounced variations in the stresses and strains occur. The most familiar examples of such regions of steep

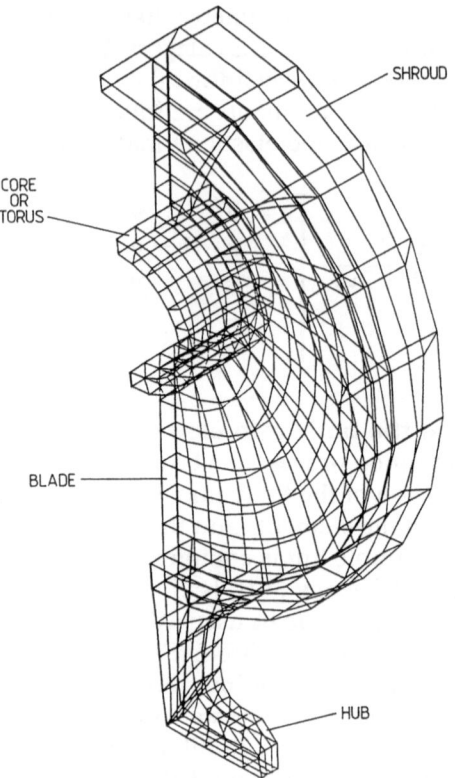

CORE
OR
TORUS

SHROUD

BLADE

HUB

FIG. 4.9. Three-dimensional finite element modelling of a sector of an impeller of a fluid coupling using 20-noded parabolic solid elements.

gradients of the variables are locations where stress concentrations exist. In this case, a finer mesh is needed at and near the geometrical discontinuity in order to provide a satisfactory solution to the problem. A typical example of this is a plate with a central hole under uniaxial tension in which the size of the elements near the hole should be very small compared with those farther away. In view of the symmetric nature of both geometry and the externally applied loads, only one quadrant can be used to model the situation. The symmetry conditions, however, must be incorporated in the solution procedure as depicted in Fig. 4.10 by the condition imposed upon the u and v displacements.

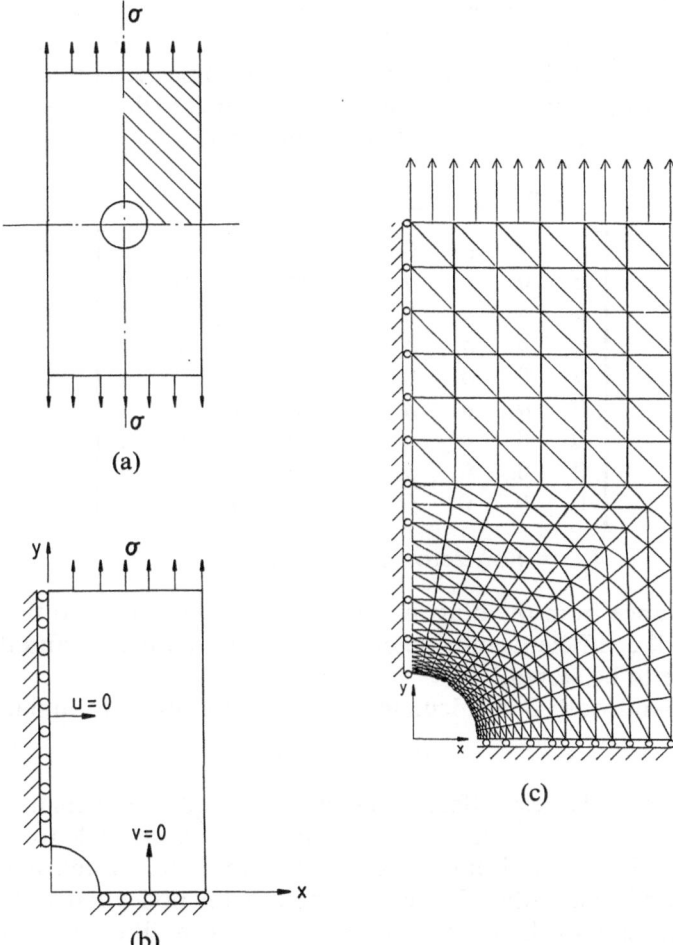

FIG. 4.10. A plate with a central hole under uniaxial tension: (a) definition of the problem; (b) one quadrant of the plate with central hole, showing imposed restraints and applied load; (c) discretised quadrant showing mesh used, applied load and imposed restraints.

4.4.4 Labelling of nodes
The labelling of the nodes (assigning a number) influences the computational efficiency associated with obtaining a solution. The set of

linear equations which arises when using the finite element method has a large number of coefficients which are zero. A listing of the equations would show that all the non-zero values and some zero values fall between two lines which can be constructed parallel to the main diagonal, as shown by the following system of equations:

$$
\xleftarrow{\hspace{1cm}} \text{Bandwidth} \xrightarrow{\hspace{1cm}}
$$

$$
\begin{bmatrix}
a & a & a & 0 & a & 0 & 0 & 0 & 0 \\
a & 0 & a & a & a & a & 0 & 0 & 0 \\
a & a & a & 0 & a & a & a & 0 & 0 \\
a & a & a & a & 0 & a & a & a & 0 \\
a & 0 & a & 0 & a & a & a & 0 & a \\
0 & a & a & a & a & 0 & a & a & a \\
0 & 0 & a & a & a & a & a & a & a \\
0 & 0 & 0 & a & a & a & 0 & a & a \\
0 & 0 & 0 & 0 & a & a & a & a & a
\end{bmatrix}
$$

(a = non-zero coefficient)

The distance from the diagonal to the dotted line is called the bandwidth. All coefficients outside of the bandwidth are zero, and they do not have to be stored.

The bandwidth B is calculated using the following formula:

$$B = (N + 1)D \tag{4.36}$$

where N is the largest difference between the node numbers in a single element and D is the number of unknowns at each node (degrees of freedom). The minimisation of B depends upon the minimisation of N which can be partially achieved by labelling the nodes across the shortest dimension of the body. Examples of the two different numbering systems are shown in Fig. 4.11. The value of B for the first numbering system is 9, whereas B for the second system is 6. By the criterion of eqn (4.36), the second system clearly minimises the bandwidth for the mesh in question [4.2].

4.4.5 Effect of mesh refinement and higher order elements upon accuracy of solution

The accuracy of the finite element solution can be improved by (i) refining the mesh and (ii) selecting higher order interpolation functions. In (i), the solution relies on an increased number of piecewise displace-

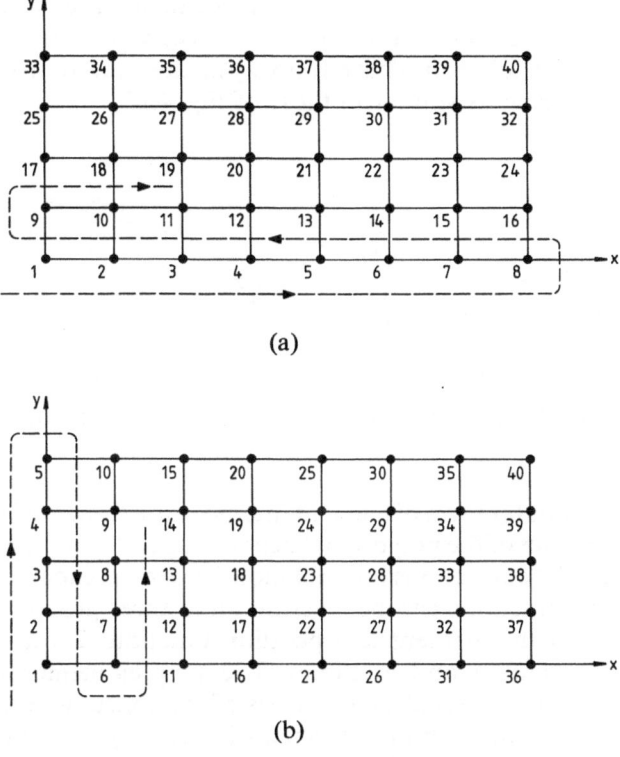

FIG. 4.11. Numbering of nodes to reduce bandwidth.

ment models of simple form to represent a complex exact solution. In (ii), the solution depends upon the selection of an improved interpolation function.

An increase in the number of elements generally means more accurate results. However, for any given problem, there will be a certain number of elements beyond which the accuracy of the solution cannot be improved significantly. Another characteristic of the discretisation that affects a finite element solution is the aspect ratio of the elements. The aspect ratio describes the shape of the element in the assemblage. For two-dimensional elements, this parameter is conveniently defined as the ratio of the largest dimension of the element to the smallest dimension. The optimum aspect ratio at any location within the discretised body depends largely upon the difference in rate of change of displacement in different directions. If the displacement varies at about the same rate in each direction, the closer the aspect ratio to unity the better the quality of the solution. This effectively means that we wish to avoid long narrow elements similar to those of Fig. 4.12.

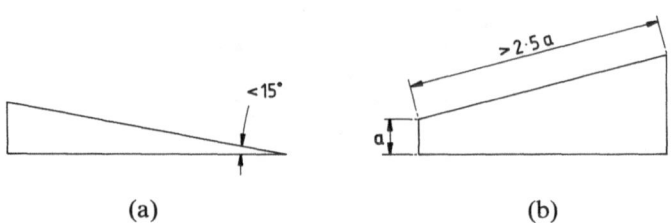

(a)   (b)

FIG. **4.12.** Poor element proportions for triangular and quadrilateral elements.

The most popular form of element interpolation function is the polynomial. The order of the polynomial depends upon the number of items known about the continuous interpolation function at each nodal point. Interpolation functions can be classified into three groups according to the order of the element interpolation function. These groupings are simplex, complex and multiplex. The simplex elements have an approximating polynomial that consists of the constant term plus the linear terms. The number of coefficients in the polynomial is equal to the dimension of the co-ordinate space plus one. The polynomial

$$\phi = a + bx + cy \qquad (4.37)$$

is the simplex function for the two-dimensional triangular element. The polynomial is linear in x and y and contains three coefficients because the triangle has three nodes.

The complex elements use a polynomial function consisting of the constant and linear terms plus second, third and higher-order terms as they are needed. The complex elements can have the same shapes as the simplex elements, but they have additional boundary nodes and may also have internal nodes. An interpolating polynomial for a two-dimensional complex triangular element is

$$\phi = a + bx + cy + dx^2 + exy + fy^2 \qquad (4.38)$$

This equation has six coefficients; therefore, the element must have six nodes.

The multiplex elements also use polynomials containing higher order terms, but the element boundaries must be parallel to the co-ordinate axes to achieve inter-element continuity. The element boundaries of the simplex and complex elements are not subjected to this restriction.

Experience indicates that fewer higher-order elements are needed to obtain the same degree of accuracy in the final answers. The use of higher order elements does not, however, always lead to a reduction in the total computation time. Numerical integration techniques are required to obtain the element matrices. These techniques can involve a large number of calculations, increasing the total computation time per element. Any additional expense for computation time, however, is probably offset by the cost decrease in the data preparation process.

The use of higher-order elements generally produces more accurate results in those applications where the gradient cannot be properly approximated by a set of constant values.

The question that concerns engineers is which of the two methods for improving accuracy is the most economical. Unfortunately, this question has not been answered definitely. There are trade-offs in time and cost as well as accuracy that must be evaluated if the question is to be answered properly. In the final analysis, what should govern the selection of an approach is the degree of accuracy obtained per unit cost.

4.5 INTERPOLATION POLYNOMIAL FUNCTIONS

From the above discussions, it is evident that the choice of suitable interpolation functions represents an important part of any finite element analysis. The appropriate choice of interpolation functions will lead to elements of high accuracy and with converging characteristics. Conversely, wrongly chosen interpolation functions may lead to poor and/or inaccurate results.

Generally, an interpolation function is either given as:

(i) a simple polynomial with undetermined coefficients which are subsequently transformed to become the relevant nodal displacements, or

(ii) directly in terms of shape functions, which are usually osculatory polynomials [4.9] having zero values at all other nodes of an element, but unit values for the displacement or its partial derivatives at the node under consideration.

Thus, the displacement field (u, v) in a two-dimensional element can either be given as

$$\left.\begin{aligned} u(x, y) &= a_1 + a_2 x + a_3 y + \ldots \\ v(x, y) &= b_1 + b_2 x + b_3 y + \ldots \end{aligned}\right] \qquad . \qquad (4.39)$$

according to (i) above, or

$$\left.\begin{aligned} u(x, y) &= u_1 N_1 + u_2 N_2 + u_3 N_3 + \ldots \\ v(x, y) &= v_1 N_1 + v_2 N_2 + v_3 N_3 + \ldots \end{aligned}\right] \qquad (4.40)$$

according to (ii). In the above expressions $a_1, a_2, a_3 \ldots b_1, b_2, b_3$, etc., are undetermined polynomial constants, N_1, N_2, N_3, etc., are shape functions and $u_1, u_2, u_3, \ldots v_1, v_2, v_3$, etc., are the unknown nodal displacements.

The detailed analysis of displacement or interpolation functions can be found in any theoretical finite element text. However, we present here a summary of the pertinent points which must be taken into account in deciding upon the order of the polynomial in a polynomial type of interpolation function. These are:

(i)   the interpolation function must satisfy the convergence requirements [4.2],

(ii) the number of generalised co-ordinates should be equal to the number of the nodal degrees of freedom of the element, and

(iii) the pattern of variation of the field variable (stress, strain, temperature, etc.) should be independent of the local co-ordinate system. This property is known as geometric invariance or spatial isotropy. In order to achieve this, the polynomial should contain terms which do not violate symmetry.

At this stage, it is appropriate to become familiar with concepts used in the construction of displacement or interpolation functions.

4.5.1 The Pascal triangle

For displacement functions of two- and three-dimensional elements given in terms of simple polynomials, the Pascal triangle of Fig. 4.13 is a useful aid for determining the combination of terms which should be used. Thus in the case of a two-dimensional simplex element, the polynomial should include a constant term plus terms containing x and y and not only one of them. Table 4.1 provides a few examples on the use of Pascal triangle in the development of the appropriate interpolation functions for two-dimensional elements.

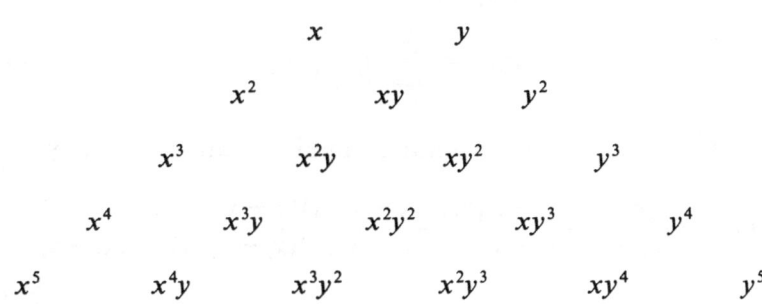

$$1$$

$$x \qquad\qquad y$$

$$x^2 \qquad xy \qquad y^2$$

$$x^3 \qquad x^2y \qquad xy^2 \qquad y^3$$

$$x^4 \qquad x^3y \qquad x^2y^2 \qquad xy^3 \qquad y^4$$

$$x^5 \qquad x^4y \qquad x^3y^2 \qquad x^2y^3 \qquad xy^4 \qquad y^5$$

FIG. 4.13. Pascal triangle.

Table 4.1
Typical Two-Dimensional Finite Elements Using Pascal Triangle [4.9]

Pascal Triangle	Element Type	Comments
1 x y x^2 xy y^2	Constant strain triangle	Linear variation of u and v $u \rightarrow 3$ DOF $v \rightarrow 3$ DOF
1 x y x^2 xy y^2	Plane stress rectangle	Linear variation of u and v along edges of element. xy term degenerates to either x or y along the appropriate edge. $u \rightarrow 4$ DOF $v \rightarrow 4$ DOF
1 x y x^2 xy y^2 x^3 x^2y xy^2 y^3	Parabolic plane stress rectangle with corner and mid-side nodes	Parabolic variation of u and v along edges. x^3 and y^3 are excluded. $u \rightarrow 8$ DOF $v \rightarrow 8$ DOF

4.5.2 Lagrange polynomials

These are often used for the construction of shape functions for elements in which only function values but not derivatives are specified at the nodes. The basic form of Lagrange's polynomial in a single co-ordinate system with n nodes is

$$u(x) = \sum_{i=0}^{n} \mathscr{L}_i^n (x) u_i \qquad (4.41)$$

where $\mathscr{L}_i^n (x)$ is known as Lagrange multiplier and is given by

$$\mathscr{L}_i^n (x) = \frac{(x - x_0)(x - x_1)\ldots(x - x_{i-1})(x - x_{i+1})\ldots(x - x_n)}{(x_i - x_0)(x_i - x_1)\ldots(x_i - x_{i-1})(x_i - x_{i+1})\ldots(x_i - x_n)}$$

$$(4.42)$$

It can be seen that $\mathscr{L}_i^n (x)$ is an nth degree polynomial as it is given by the product of n linear factors. It is also obvious that $\mathscr{L}_i^n (x)$ possesses the properties

$$\mathscr{L}_i^n (x_k) = \begin{cases} 0 & k \neq i \\ 1 & k = i \end{cases} \qquad (4.43)$$

and thus fits in with the definition of a shape function.

It is obviously possible to use Lagrange's formulation to shape functions involving two and three co-ordinates, e.g. the shape functions for a two-dimensional problem would be

$$u(x, y) = \sum_{i=0}^{n} \sum_{j=0}^{m} \mathscr{L}_i^n (x) \, \mathscr{L}_j^m (y) \, u_{ij} \qquad (4.44)$$

where n and m are the number of subdivisions in the x and y directions, respectively. In fact, the shape functions given by (4.44) form the basis of a family of plane stress elements [4.9].

As an extension to Lagrange polynomials, the Hermitian polynomials were developed not only to agree in value with a given function at specified locations, but their derivatives will also match with the derivatives of the given functions up to any given order at any particular location. These functions are commonly used in the analysis of bending of finite strips and plates.

It is worth noting that all the previous discussions are concerned with Cartesian co-ordinates (x, y), and the examples given are valid only for rectangular elements. For general quadrilateral elements similar functions can be used, but they should be written in terms of intrinsic or curvilinear co-ordinates (ξ, η). A detailed discussion of this will be provided in the following section. The Cartesian co-ordinates are also not suitable for triangular elements, and in this case a special type of co-ordinate system known as area co-ordinates is adopted.

4.6 ISOPARAMETRIC FINITE ELEMENT REPRESENTATION

4.6.1 Intrinsic co-ordinate system

A local co-ordinate system (ξ, η) is one which is defined for a particular element and not necessarily for the entire body or structure. The co-ordinate system for the entire body is known as the global system (x, y). An intrinsic (natural) co-ordinate system is a local system which permits the identification of a point within the element by a set of dimensionless numbers whose magnitudes never exceed unity (Fig. 4.14). These systems are usually so arranged that some of the intrinsic co-ordinates have unit magnitudes at specific nodal points. Furthermore, such a structure simplifies the formulation of finite element codes and facilitates the numerical integration necessary for the determination of element stiffness matrices.

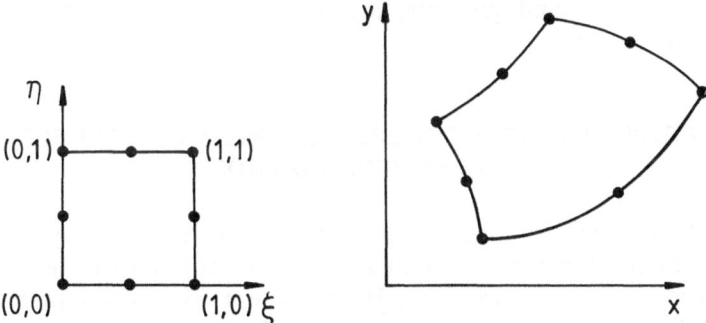

FIG. 4.14. Isoparametric mapping.

4.6.2 Isoparametric elements

The displacement field (u, v) for a two-dimensional element of general shape can be given by

$$u(x, y) = \sum_{i=1}^{n} u_i N_i$$
$$= u_1 N_1 + u_2 N_2 + u_3 N_3 \ldots u_n N_n \qquad (4.45)$$

and

$$
\left.
\begin{aligned}
v(x,y) &= \sum_{i=1}^{n} v_i N_i \\
&= v_1 N_1 + v_2 N_2 + v_3 N_3 \ldots v_n N_n
\end{aligned}
\right\}
\tag{4.46}
$$

where N_i is a function of the intrinsic co-ordinates (ξ, η) and n is the number of nodes. The co-ordinates (x,y) inside the element domain can also be described in a similar manner by the following expressions:

$$
\left.
\begin{aligned}
x &= \sum_{i=1}^{n} x_i M_i \\
&= x_1 M_1 + x_2 M_2 + x_3 M_3 \ldots x_n M_n
\end{aligned}
\right\}
\tag{4.47}
$$

and

$$
\left.
\begin{aligned}
y &= \sum_{i=1}^{n} y_i M_i \\
&= y_1 M_1 + y_2 M_2 + y_3 M_3 \ldots y_n M_n
\end{aligned}
\right\}
\tag{4.48}
$$

in which M_i is also a function of ξ and η. For the particular case in which N_i and M_i are identical, i.e. the shape functions defining the displacement fields and geometry are the same, the element is said to be isoparametric.

4.7 DERIVATION OF A GENERAL EXPRESSION FOR ELEMENT STIFFNESS MATRIX

One of the most important items in finite element analysis is the development of general element stiffness matrices for two- and three-dimensional elasticity problems. In the present work, we adopt the principle of minimum potential energy for deriving the equilibrium equations. Since the nodal degrees of freedom are treated as unknown in the present displacement formulation, the potential energy U has to be first expressed in terms of nodal degrees of freedom. The necessary equilibrium can be obtained by setting the first partial derivatives of U with respect to each of the nodal degrees of freedom equal to zero. The following is an outline of the required steps for the derivation of the element stiffness matrix for three-dimensional elasticity problems of the simplex type. This approach is not directly applicable to the isoparametric elements discussed earlier.

(i) Define the displacement and force vectors for a general element as follows:

$$\{\delta\}_{3n \times 1} = \{u_1 v_1 w_1 \; u_2 v_2 w_2 \ldots u_n v_n w_n\}^{\mathrm{T}} \qquad (4.49)$$

$$\{F\}_{3n \times 1} = \{F_{x1} F_{y1} F_{z1} \; F_{x2} F_{y2} F_{z2} \ldots F_{xn} F_{yn} F_{zn}\}^{\mathrm{T}} \qquad (4.50)$$

where n is the number of nodes in the desired element.

(ii) Express the displacement at any point within the element in terms of the nodal displacements, i.e.

$$
\left.
\begin{aligned}
u(x,y,z) &= \sum_{i=1}^{n} u_i N_i \\
v(x,y,z) &= \sum_{i=1}^{n} v_i N_i \\
w(x,y,z) &= \sum_{i=1}^{n} w_i N_i
\end{aligned}
\right\}
\qquad (4.51)
$$

where u_i, v_i and w_i are the nodal displacements of the ith node and N_i is the corresponding element shape function.

(iii) Express the strain components at any point within the element in terms of the nodal displacements:

$$
\{\varepsilon\} =
\begin{Bmatrix}
\varepsilon_{xx} \\
\varepsilon_{yy} \\
\varepsilon_{zz} \\
\gamma_{xy} \\
\gamma_{yz} \\
\gamma_{zx}
\end{Bmatrix}
=
\begin{Bmatrix}
\dfrac{\partial u}{\partial x} \\[2mm]
\dfrac{\partial v}{\partial y} \\[2mm]
\dfrac{\partial w}{\partial z} \\[2mm]
\dfrac{\partial u}{\partial y} + \dfrac{\partial v}{\partial x} \\[2mm]
\dfrac{\partial v}{\partial z} + \dfrac{\partial w}{\partial y} \\[2mm]
\dfrac{\partial w}{\partial x} + \dfrac{\partial u}{\partial z}
\end{Bmatrix}
=
$$

$$= \begin{bmatrix} \dfrac{\partial N_1}{\partial x} & 0 & 0 & \dfrac{\partial N_2}{\partial x} & 0 & 0 & \cdots & \dfrac{\partial N_n}{\partial x} & 0 & 0 \\[2mm] 0 & \dfrac{\partial N_1}{\partial y} & 0 & 0 & \dfrac{\partial N_2}{\partial y} & 0 & \cdots & 0 & \dfrac{\partial N_n}{\partial y} & 0 \\[2mm] 0 & 0 & \dfrac{\partial N_1}{\partial z} & 0 & 0 & \dfrac{\partial N_2}{\partial z} & \cdots & 0 & 0 & \dfrac{\partial N_n}{\partial z} \\[2mm] \dfrac{\partial N_1}{\partial y} & \dfrac{\partial N_1}{\partial x} & 0 & \dfrac{\partial N_2}{\partial y} & \dfrac{\partial N_2}{\partial x} & 0 & \cdots & \dfrac{\partial N_n}{\partial y} & \dfrac{\partial N_n}{\partial x} & 0 \\[2mm] 0 & \dfrac{\partial N_1}{\partial z} & \dfrac{\partial N_1}{\partial y} & 0 & \dfrac{\partial N_2}{\partial z} & \dfrac{\partial N_2}{\partial y} & \cdots & 0 & \dfrac{\partial N_n}{\partial z} & \dfrac{\partial N_n}{\partial y} \\[2mm] \dfrac{\partial N_1}{\partial z} & 0 & \dfrac{\partial N_1}{\partial x} & \dfrac{\partial N_2}{\partial z} & 0 & \dfrac{\partial N_2}{\partial x} & \cdots & \dfrac{\partial N_n}{\partial z} & 0 & \dfrac{\partial N_n}{\partial x} \end{bmatrix} \begin{Bmatrix} u_1 \\ v_1 \\ w_1 \\ u_2 \\ v_2 \\ w_2 \\ - \\ - \\ - \\ - \\ u_n \\ v_n \\ w_n \end{Bmatrix}$$

$$\text{(4.52)}$$

i.e. $\qquad \{\varepsilon\}_{6 \times 1} = [B]_{6 \times 3n} \cdot \{\delta\}_{3n \times 1}$ $\qquad$ (4.53)

(iv) Express the stress components at a general point in the element in terms of the nodal displacements. Assuming the material is perfectly elastic, homogeneous and isotropic, then

$$\{\sigma\}_{6 \times 1} = [D]_{6 \times 6} \cdot \{\varepsilon\}_{6 \times 1} \qquad \text{(4.54)}$$

or alternatively

$$\{\sigma\} = [D] \cdot [B] \cdot \{\delta\} \qquad \text{(4.55)}$$

where $[D]$ is the elasticity matrix, and is defined by eqns(4.15).

(v) The element stiffness matrices and load vectors are to be derived from the application of the principle of minimum potential energy to the above system. For this, the potential energy U is obtained by considering the strain energy and the work done by the external forces as follows:

$$U = (1/2) \int_{V^{(e)}} \{\varepsilon\}^T \{\sigma\} dV - \{\delta\}^T \cdot \{F\}$$

$$= (1/2) \int_{V^{(e)}} \{\delta\}^T [B]^T [D] [B] \{\delta\} \, dV - \{\delta\}^T \cdot \{F\} \qquad \text{(4.56)}$$

The minimisation of the above equation with respect to the displacement vector $\{\delta\}$ will yield the following expression for the element stiffness matrix for general three-dimensional elastic problems:

$$\left(\int_{V^{(e)}} [B]^\mathrm{T} [D] [B] \, \mathrm{d}V\right) \{\delta\} = \{F\} \tag{4.57}$$

or

$$[K^{(e)}]_{3n \times 3n} \cdot \{\delta\}_{3n \times 1} = \{F\}_{3n \times 1} \tag{4.58}$$

where $[K^{(e)}]$ is known as the element stiffness matrix and is given by

$$[K^{(e)}] = \int_{V^{(e)}} [B]^\mathrm{T} [D] [B] \, \mathrm{d}V \tag{4.59}$$

for three-dimensional problems, and

$$[K^{(e)}] = t \int_{A^{(e)}} [B]^\mathrm{T} [D] [B] \, \mathrm{d}A \tag{4.60}$$

for two-dimensional problems.

4.7.1 Isoparametric elements: stiffness formulation

The $[B]$ matrix given in equation (4.60) for two-dimensional problems expresses the strains in terms of the nodal displacements and therefore contains first derivatives of the shape functions with respect to global x and y axes. However, the shape functions for isoparametric elements are defined in terms of the intrinsic co-ordinates ξ and η, and therefore cannot be differentiated directly with respect to the global x and y axes. (Typical shape functions are provided in Appendix 4.1).

In order to overcome this difficulty it is necessary to obtain a relationship between the derivatives of the two sets of co-ordinates as follows:

$$\left.\begin{aligned} \frac{\partial N}{\partial \xi} &= \frac{\partial N}{\partial x}\frac{\partial x}{\partial \xi} + \frac{\partial N}{\partial y}\frac{\partial y}{\partial \xi} \\ \frac{\partial N}{\partial \eta} &= \frac{\partial N}{\partial x}\frac{\partial x}{\partial \eta} + \frac{\partial N}{\partial y}\frac{\partial y}{\partial \eta} \end{aligned}\right\} \tag{4.61}$$

which in matrix form can be written as:

$$
\left\{ \begin{array}{c} \dfrac{\partial N}{\partial \xi} \\[2mm] \dfrac{\partial N}{\partial \eta} \end{array} \right\}
=
\begin{bmatrix} \dfrac{\partial x}{\partial \xi} & \dfrac{\partial y}{\partial \xi} \\[2mm] \dfrac{\partial x}{\partial \eta} & \dfrac{\partial y}{\partial \eta} \end{bmatrix}
\left\{ \begin{array}{c} \dfrac{\partial N}{\partial x} \\[2mm] \dfrac{\partial N}{\partial y} \end{array} \right\}
=
$$

$$
[J] \cdot \left\{ \begin{array}{c} \dfrac{\partial N}{\partial x} \\[2mm] \dfrac{\partial N}{\partial y} \end{array} \right\} \tag{4.62}
$$

The matrix $[J]$ relating the derivatives of the two systems is known as the Jacobian matrix and its coefficients can be obtained by differentiating expressions (4.47) and (4.48) with respect to the ξ and η co-ordinates. To complete the transformation between the two systems, it is also necessary to express the elemental area dA of equation (4.60) in terms of ξ and η as follows:

$$
dA = \det [J] \, d\xi d\eta \tag{4.63}
$$

It follows that the stiffness matrix for two-dimensional isoparametric elements can be written as:

$$
[K^{(e)}] = t \int_0^1 \int_0^1 [B]^T [D] [B] \det [J] \, d\xi d\eta \tag{4.64}
$$

in which $[B]^T \cdot [D] \cdot [B]$ is only a function of the intrinsic co-ordinates ξ and η. It is worth noting that in order to obtain $\partial N/\partial x$ and $\partial N/\partial y$ it is necessary to invert $[J]$. This inverse exists if there is no excessive distortion of the element.

The above integrals are commonly evaluated numerically using Gaussian quadrature method. The details can be found in the appropriate literature on finite elements and numerical analysis. It is worth remembering that in order to effect a solution to the entire system, one must apply (4.64) to that entire system. This will enable the determination of the equilibrium equation for the body or structure.

4.8 COMPUTER IMPLEMENTATION OF FINITE ELEMENT

One of the advantages of the finite element method is that many of the solution steps are common to most application areas. The procedure for solving, e.g. problems in heat transfer and fluid flow, involve many of the same steps found in structural stress and dynamic response analyses. Figure 4.15 shows a simplified computer flow diagram for an elastic finite element code, the major steps of which will be discussed in general terms rather than with respect to any specific example.

Before any analysis can be made, it is necessary for the designer to make two important decisions: (i) the type of element and (ii) the mesh configuration to be used. Both of these decisions are dictated by the geometrical features of the component, the applied loads, the boundary conditions, stress gradients and accuracy in relation to cost.

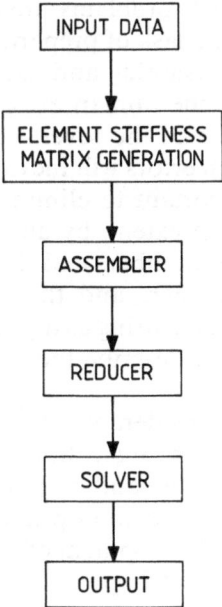

FIG. **4.15.** A general computer flow diagram for a finite element program.

For a number of elements, and in particular the constant strain triangular elements, the accuracy of the results deteriorates rapidly with increased aspect ratio and, in general, long and narrow elements should be avoided (Fig. 4.12). It is also worth noting that there are no specific rules for establishing the optimum mesh for a new design, except that:

(i) common boundaries must have the same number of divisions and ratios to ensure compatibility, and

(ii) the design must be modelled into groups of simpler geometric blocks with known shape functions.

In these situations however, experience plays an important role in the decision making process. One common approach relies on the use of successfully refined meshes until acceptable convergence has been achieved.

4.8.1 Automatic mesh generation

Most practical engineering problems involve a large number of elements and nodes and the task of preparing the input data becomes extremely lengthy, time consuming and tedious. Furthermore, during the preparation of data some human errors may be introduced and remain undetected in spite of the several checks which are usually made. The presence of such errors will inevitably bring about incorrect results. It is therefore important to eliminate such data errors. This can be achieved, to a large extent, by automating the discretisation process. In this case, nodal numbers and their associated co-ordinates together with element numbers and their definitions are prepared automatically by the computer, using as input data the minimal amount of information necessary to describe the component and the desired mesh divisions.

In the past few years, considerable efforts have been expended in developing mesh generation algorithms and thus improving the efficiency of data preparation and minimising data errors. (These algorithms can be found in the appropriate literature). This has also been supplemented by the development of interactive mesh generation facilities, detailed in the next Chapter.

4.8.2 Input data

After having decided on element type and mesh division, it is now possible to prepare the data for input to the program. Such data includes all information concerning component geometry, element types, node numbering and co-ordinates, material mechanical and physical properties, applied loads and imposed restraints.

In most finite element programs a data echo is included and the input data are printed as a matter of course. The inclusion of such a data echo will allow the checking of input data and will enable the identification of the studied case for future reference.

4.8.3 Element stiffness matrix generation and assembly

All finite elements computer programs require information regarding the number of equations, the number of elements and the bandwidth. Information about the number of equations is needed so that the global (overall) stiffness matrix and the global force vector can be initialised to zero. Initialisation is necessary because the matrices are built by a summation process. The initialisation process is followed by a 'DO' loop through the element. At this instant, the element data (described earlier) are read and the appropriate element matrices are generated and then inserted in their appropriate position in the global matrix. Obviously, the computation of element stiffness matrices requires information about the material mechanical and physical properties. These properties include Young's modulus, Poisson's ratio, coefficient of thermal conductivity, convection heat transfer coefficient, density, etc. The evaluation of the Jacobian matrix along with its inverse and determinant are also needed. The partial derivatives needed to evaluate the Jacobian matrix for the appropriate elements are computed in a subroutine which has equations for both the shape functions and their derivatives.

It must be pointed out however that in a typical finite element using numerically integrated isoparametric elements and an advanced solution subroutine such as 'the front solver', the computer time spent in calculating the element stiffness can vary from 25 to 85% of the total time required for a complete analysis. Therefore, the algorithm used to compute the element stiffness has a significant effect on the overall efficiency of the program. With the algorithms in common use, the computer time required to calculate an element stiffness matrix varies by a factor of ten between the best and the least efficient.

The stiffness relationship for an element has been given by equation (4.58) as

$$[K^{(e)}] \{\delta\} = \{F\} \cdot$$

For a 4-noded rectangular element, the above expression can be written as:

$$\begin{bmatrix} K_{ii} & K_{ij} & K_{il} & K_{im} \\ K_{ji} & K_{jj} & K_{jl} & K_{jm} \\ K_{li} & K_{lj} & K_{ll} & K_{lm} \\ K_{mi} & K_{mj} & K_{ml} & K_{mm} \end{bmatrix} \begin{Bmatrix} \delta_i \\ \delta_j \\ \delta_l \\ \delta_m \end{Bmatrix} = \begin{Bmatrix} P_i \\ P_j \\ P_l \\ P_m \end{Bmatrix} \tag{4.65}$$

in which all the terms are in fact submatrices.

Since a component is made up of many elements, it follows that the component global stiffness matrix is also made up of a corresponding number of element stiffness matrices, thus

$$[K]\{\delta\} = \{F\}$$

where $[K]$ is the global stiffness matrix $\left(= \sum_{e=1}^{n} [K^{(e)}], \right.$ where n is the number of elements $\left.\right)$,

$\{\delta\}$ is the global vector of nodal displacements, and
$\{F\}$ is the global force vector with a number of load cases.

It is worth noting that the nodal force vector consists of quantities that are associated with a specific node rather than a specific element. In general, most of these quantities are associated with boundary nodes such as concentrated or distributed forces in solid mechanics and heat losses in heat transfer problems. The values and location of these forces will become obvious in the solving of the problem, and their determination will be considered for the different case studies discussed in the next Chapter. The process of assembly of element stiffness matrices has already been explained in section 4.2.

4.8.4 Solution of simultaneous equations
The resulting system of equations which is characterised by

$$[K]\{\delta\} = \{F\}$$

and obtained by summation over all the elements, must be modified whenever some of the values in the displacement vector $\{\delta\}$ are known. This is more the rule rather than the exception.

Most field problems have some of the boundary values specified, and all elasticity problems must have some of the displacements specified in order to eliminate rigid body degrees of freedom. As a matter of fact, the global stiffness matrix $[K]$ discussed earlier is singular until some of the displacements have been specified.

The modification of the system of equations is followed by the solution of these equations for the nodal unknowns. There are several procedures for accomplishing this solution. These include the direct solvers using Gaussian elimination, Gauss–Jordan elimination or Choleski's factorisation and the iterative solvers using the relaxation approach, Gauss–Seidel or the conjuget–gradient methods. The techniques adopted by these methods as well as merits and limitations can be further investigated in general numerical analysis books.

The resulting system of equations is special because it is banded and the diagonal terms are usually positive and dominate their respective row and column. This allows many of the above mentioned solvers to be modified for increased efficiency and compactness.

The solution of the system of equations is followed by output of the nodal displacement. Once the displacements have been obtained, it is possible to compute the stresses of each element in turn by using equation (4.55), viz.

$$\{\sigma\} = [D] [B] \{\delta\}$$

4.9 CONCLUSIONS

There is no discussion in this Chapter concerning the mathematical foundations of the finite element method. This topic has already received considerable attention from mathematicians in the past. Several books that discuss the mathematical aspects of the method are available in the literature; some are obviously far superior to others. However, it is the author's belief that this text provides the basic foundation on which the designer can build a sound and in-depth knowledge of this method.

Undoubtedly, the finite element method is here to stay; it is a powerful technique for obtaining numerical solutions to practical engineering problems. The designer of the future will have to learn the basic concepts of the method and remain current with them.

REFERENCES

[4.1]   O. C. Zienkiewicz, *The Finite Element Method in Engineering Science,* McGraw Hill, London, 1977.

[4.2]   C. S. Desai and J. F. Abel, *Introduction to the Finite Element Method,* Van Nostrand Reinhold Company, New York, 1972.

[4.3]   B. Nath, *Fundamentals of Finite Elements for Engineers,* The Athlone Press of University of London, 1974.

[4.4]   R. T. Fenner, *Finite Element Methods for Engineers,* McMillan, London, 1975.

[4.5]   L. J. Segerlind, *Applied Finite Element Analysis,* John Wiley & Son, New York, 1976.

[4.6]   S. S. Rao, *The Finite Element Method in Engineering,* Pergamon Press, Oxford, 1982.

[4.7]   J. P. Den Hartog, *Advanced Strength of Materials,* McGraw Hill, New York, 1952.

[4.8]   A. P. Boresi, *Elasticity in Engineering Mechanics,* Prentice-Hall International, Englewood Cliffs, NJ, 1965.

[4.9]   Y. K. Cheung and M. F. Yeo, *A Practical Introduction to Finite Element Analysis,* Pitman Publishing Limited, London, 1979.

APPENDIX 4.1

Shape functions for a few common elements in terms of intrinsic co-ordinates

(i) Two-noded linear bar elements

$$N_1(\xi) = 1 - \xi$$
$$N_2(\xi) = \xi$$

(ii) Three-noded parabolic bar elements

$$N_1(\xi) = (1 - \xi)(1 - 2\xi)$$
$$N_2(\xi) = 4\xi(1 - \xi)$$
$$N_3(\xi) = -\xi(1 - 2\xi)$$

(iii) Four-noded linear plane elements

$$N_1(\xi, \eta) = (1 - \xi)(1 - \eta)$$

$$N_2(\xi, \eta) = \xi(1 - \eta)$$

$$N_3(\xi, \eta) = \xi\eta$$

$$N_4(\xi, \eta) = (1 - \xi)\eta$$

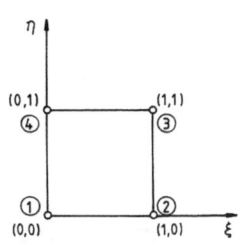

(iv) Eight-noded quadratic elements

$$N_1(\xi, \eta) = (1 - \xi)(1 - \eta)(1 - 2\xi - 2\eta)$$

$$N_2(\xi, \eta) = 4\xi(1 - \xi)(1 - \xi)$$

$$N_3(\xi, \eta) = \xi(1 - \eta)(-1 + 2\xi - 2\eta)$$

$$N_4(\xi, \eta) = 4\xi\eta(1 - \eta)$$

$$N_5(\xi, \eta) = \xi\eta(-3 + 2\xi + 2\eta)$$

$$N_6(\xi, \eta) = 4\xi(1 - \xi)\eta$$

$$N_7(\xi, \eta) = (1 - \xi)\eta(-1 - 2\xi + 2\eta)$$

$$N_8(\xi, \eta) = 4(1 - \xi)\eta(1 - \eta)$$

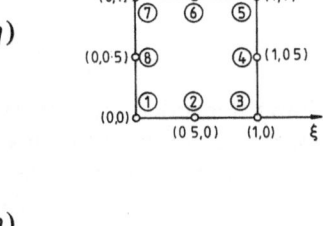

(v) Eight-noded solid brick elements

$$N_1(\xi, \eta, \zeta) = (1 - \xi)(1 - \eta)(1 - \zeta)$$

$$N_2(\xi, \eta, \zeta) = \xi(1 - \zeta)(1 - \eta)$$

$$N_3(\xi, \eta, \zeta) = \xi\eta(1 - \zeta)$$

$$N_4(\xi, \eta, \zeta) = (1 - \xi)\eta(1 - \zeta)$$

$$N_5(\xi, \eta, \zeta) = (1 - \xi)(1 - \eta)\zeta$$

$$N_6(\xi, \eta, \zeta) = \xi(1 - \eta)\zeta$$

$$N_7(\xi, \eta, \zeta) = \xi\eta\zeta$$

$$N_8(\xi, \eta, \zeta) = (1 - \xi)\eta\zeta$$

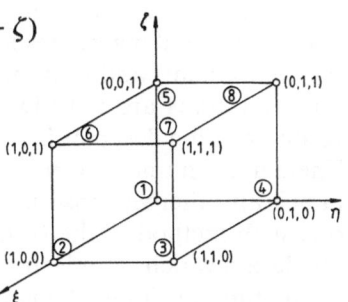

Chapter 5

Computer-Aided Analysis

5.1 INTRODUCTION

Computer-aided analysis (CAA) is the name given to the analysis and optimising parts of the design process which, together with computer-aided design and computer-integrated manufacture, form the complete design package. The benefits of integrating these approaches with computer aids are immense; they include decreased lead time, superior and efficient designs and reduced manufacturing costs.

In order to evaluate the mechanical integrity of an engineering design, some form of analysis is required. This analysis may involve static deflection, strain and stress calculations, heat transfer, fluid dynamics and dynamic response studies. Due to the complex nature of the components and/or applied mechanical and thermal loads, closed form solutions are not always possible. Faced with this, the natural recourse of the designer is to seek a numerical solution to the problem. There are a number of numerical procedures available with which to attack an otherwise insoluble problem. The finite element method is one such method. If the finite element method indicates behaviour of the design which is undesirable, the designer can modify the shape and recompute the finite element analysis for the revised design.

Until recently, only high technology industries such as aerospace and nuclear power have had the justification and resources to make use of these methods. However, the development of high speed digital computers, together with a greatly increased output of good quality general purpose finite element software, and the rapid decrease in computer hardware cost, puts CAA well within reach of many smaller companies.

In this chapter, the structural integrity and dynamic performance of the critical components of the wheel-assembly of centrifugal peening equipment and the fluid coupling are examined, and modifications to the previous designs are proposed, using a commercially available general purpose finite element package (SDRC-SUPERB analysis).

5.2 GENERAL PURPOSE FINITE ELEMENT SOFTWARE

Finite element computer programs (FE CODES) have become widely available analytical tools in all industrialised countries. Estimates place the total number of commercial finite element codes at over 500 worldwide covering a large number of disciplines. A number of computer program packages have been developed for the solution of a variety of structural and solid mechanics problems. A summary of the more widely used packages and their capabilities is provided in Table 5.1 (after Refs [5.1], [5.2] and [5.3]).

The evaluation of commercial finite element packages and the adequacy of their elements has received much attention in the past 5 years; see, for example, Refs [5.4] to [5.6]. Code selection is an important but difficult task, especially for new and inexperienced users. All of the major packages, however, work on a similar basis and in many cases use common element types. Figure 5.1 shows the most common elements used in general purpose FE software.

The continued development of computational capabilities has allowed the designer to solve succeedingly larger and computationally demanding problems. In turn, the capability to effectively perform more accurate analyses has allowed the designer to consider structural systems which, in view of their complexity or intended function, could only be examined by analysis.

One major difference between finite element packages is the amount of pre- and post-processing necessary before the analysis program can accept the data. Indeed, the data input to an FE program and the resulting analysis output can be considerable especially for 3-D problems. The designer is thus faced with a problem of data base definition and management. The obvious solution is to utilise the computer in this phase of data preparation and output display. This makes sense, since in the design/analysis cycle a great deal of time is spent in data processing. It has been estimated that data processing accounts for 80% of the cost of most analyses.

Table 5.1
Structural Mechanics Software (After Ref [5.2])

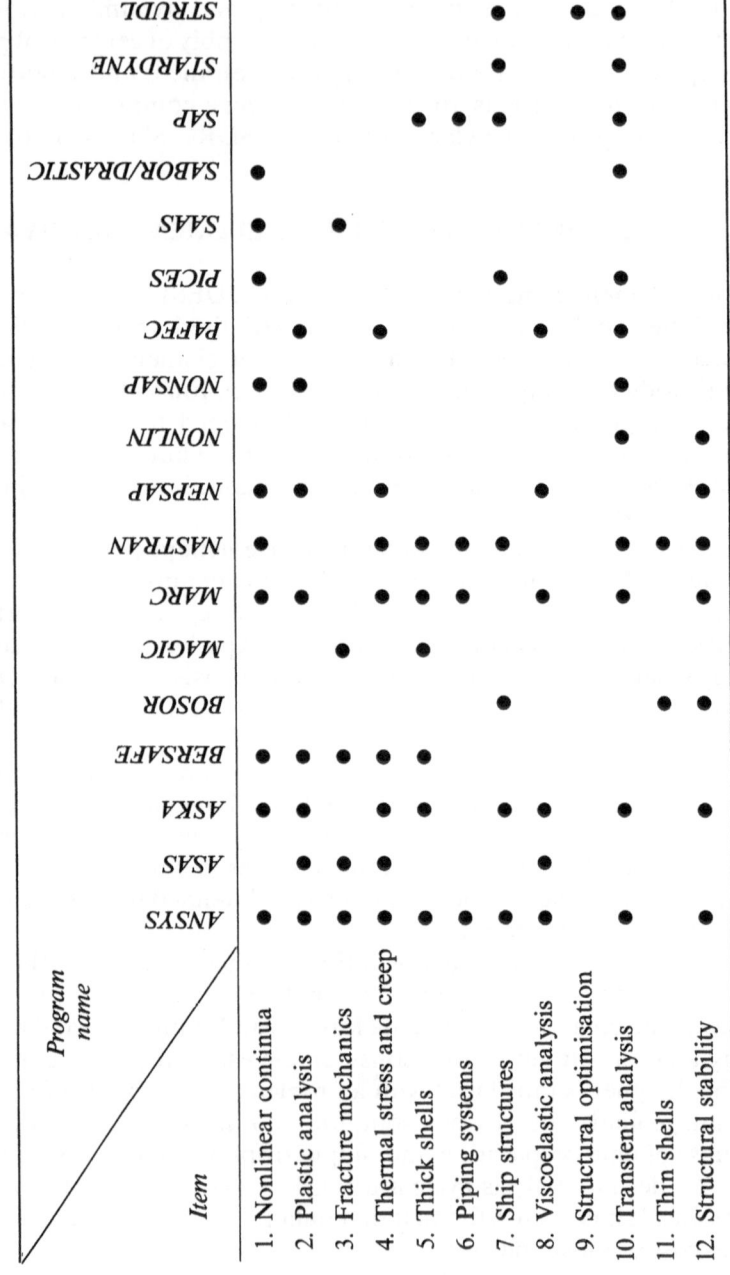

Item \ Program name	ANSYS	ASAS	ASKA	BERSAFE	BOSOR	MAGIC	MARC	NASTRAN	NEPSAP	NONLIN	NONSAP	PAFEC	PICES	SAAS	SABOR/DRASTIC	SAP	STARDYNE	STRUDL
1. Nonlinear continua	•		•	•			•	•	•		•		•	•	•			
2. Plastic analysis	•	•	•	•			•		•		•	•						
3. Fracture mechanics	•	•	•	•										•				
4. Thermal stress and creep	•	•	•	•		•	•	•	•			•				•		
5. Thick shells	•		•	•		•	•	•								•		
6. Piping systems	•						•	•										
7. Ship structures	•		•					•					•			•	•	•
8. Viscoelastic analysis	•	•			•		•											
9. Structural optimisation			•						•			•						•
10. Transient analysis	•						•	•		•								
11. Thin shells					•			•										
12. Structural stability	•		•		•		•	•	•	•		•	•		•	•	•	•

ELEMENT	DESCRIPTION	COMMENTS
LINEAR SOLIDS	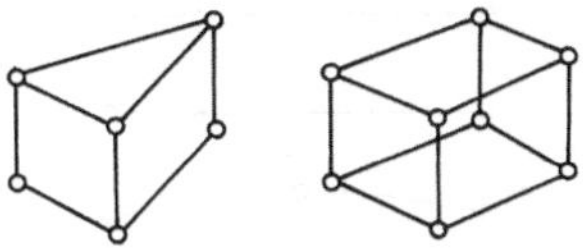	6 AND 8 NODES
PARABOLIC SOLIDS	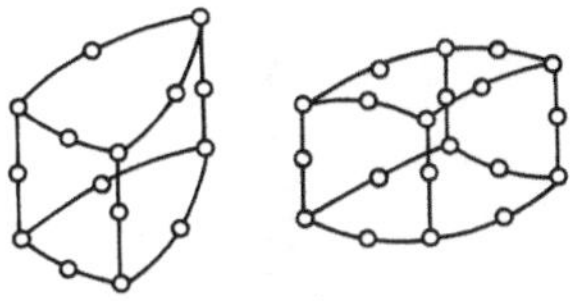	15 AND 20 NODES
SOLID TETRAHEDRA	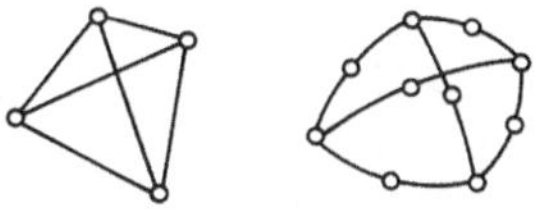	4 AND 10 NODES
LINEAR AXISYMMETRIC SOLIDS	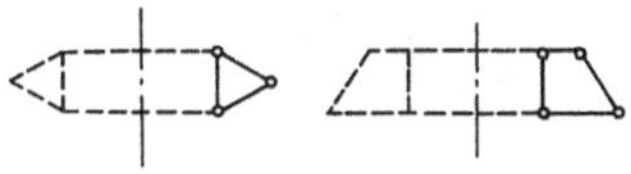	3 AND 4 NODES
PARABOLIC AXISYMMETRIC SOLIDS		6 AND 8 NODES

FIG. 5.1. Finite element library for structural analysis.

5.3 COMPUTER-AIDED SYSTEM FOR DATA PROCESSING

5.3.1 Pre- and post-processing

The user's major involvement with a computer program is during the data preparation and output evaluation phases. As discussed above, this represents a major cost in the design/analysis cycle and it is precisely in these areas that the computer can be of major assistance.

A pre-processor is used to define, verify and format an acceptable data file from a finite element mesh. This data will typically consist of mesh definition which includes geometrical features, nodes, elements,

ELEMENT	DESCRIPTION	COMMENTS
BAR	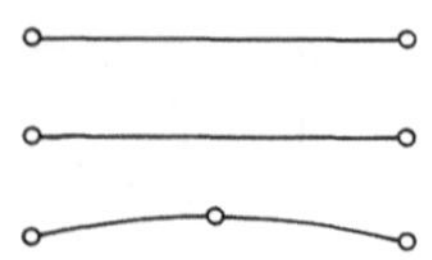	3 DOF PER NODE – AXIAL FORCE ONLY
BEAM - 2 NODED		STANDARD BEAM
BEAM - 3 NODED		REINFORCING BEAM USED WITH PARABOLIC SHELL
MASS		MASS & INERTIA PROPERTIES
SPRING TO GROUND		
LINEAR SHELL	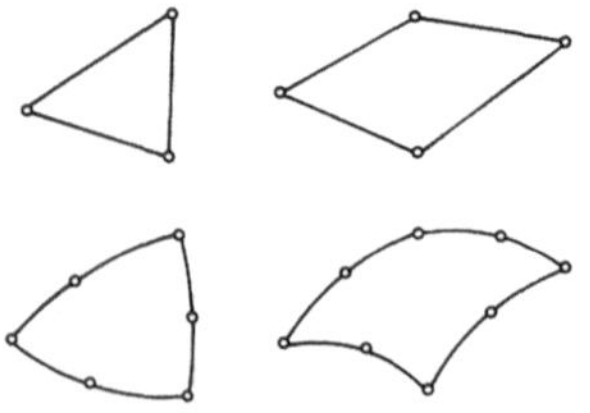	3 AND 4 NODES
PARABOLIC SHELL		6 AND 8 NODES
AXISYMMETRIC SHELL	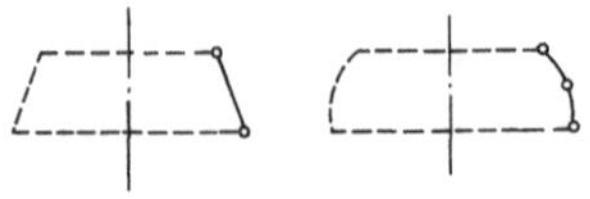	2 AND 3 NODES

FIG. 5.1. — *contd.*

material and physical properties, applied thermal and mechanical loads and imposed restraints (boundary conditions), and other additional data. Figure 5.2(a) shows how the pre-processor interfaces with the analysis software.

A post-processor, on the other hand, is used to determine the significance of the analysis results. It enables the user to interrogate the output by utilising several types of display forms, including:

(i) display analysis result contours according to specified data types such as von Mises equivalent stresses, Tresca maximum shear stresses, principal stresses, etc.,

   (ii) display and animate displaced nodes and/or elements of deformed geometry,

  (iii) display only those elements which satisfy a given criterion, e.g. displacements above a prescribed value,

  (iv) tabulate and sort analysis data for those areas in which the designer is interested.

This effective data processor may incorporate interactive graphics devices. Most of the large finite element packages now incorporate such devices to provide for automatic mesh generation and output display facilities. These processors allow the creation of models which can be processed by several different finite element packages. Figure 5.2(b) shows how the post-processor interfaces with the analysis software.

5.4 INTERACTIVE MESH GENERATION

Most pre-processors are equipped with the appropriate modules to interactively generate nodes and elements based on 2-D and 3-D surface geometries. In this work, we utilised two modules: (i) Enhanced Mesh Generation and (ii) Mapped Mesh Generation facilities of SDRC-SUPERTAB [5.7].

The Enhanced Mesh Generation (EMG) facility allows the user to automatically produce a mesh pattern by a process of logical subdivision. It divides all surfaces, regardless of shape, into progressively smaller areas until it reaches a user-specified element size. It also optimises the shape of each element based upon its aspect ratio. These meshes are particularly appropriate for irregular shapes, e.g. 3-D models with holes, and also models requiring variation in element size and concentration. These irregularities can prove more difficult for the traditional 'mapping' style mesh generators.

The Mapped Mesh Generation (MMG) facility, on the other hand, enables the user to define and generate nodes and elements on existing surfaces and volumes. Once the surface or volume for mesh generation has been selected the system will prompt the designer for the number of elements to be generated along each edge and the desired element concentration for each edge. Examples of automatic mesh generation using EMG and MMG of SDRC-SUPERTAB are provided in the next sections which deal with the computer-aided analysis aspects of the previously discussed case studies.

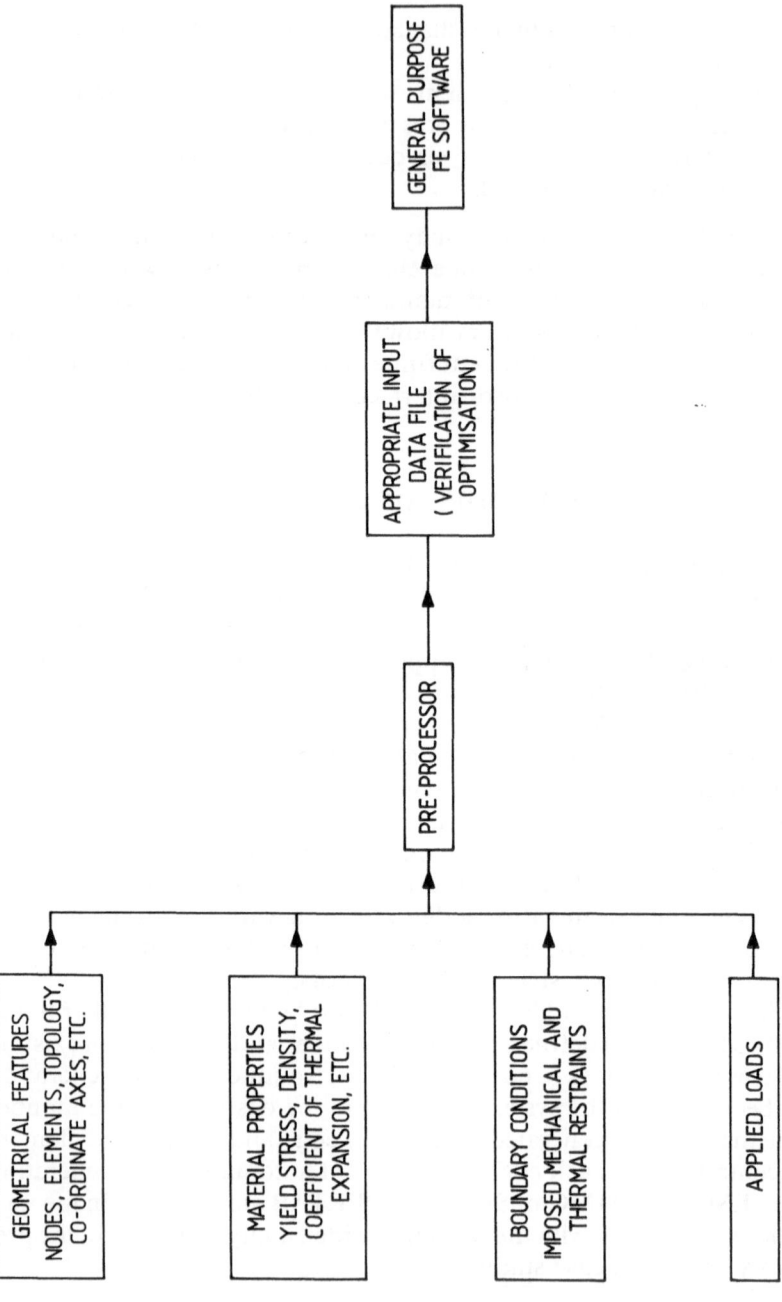

FIG. 5.2. Computer-aided system for data processing: (a) pre-processor interface.

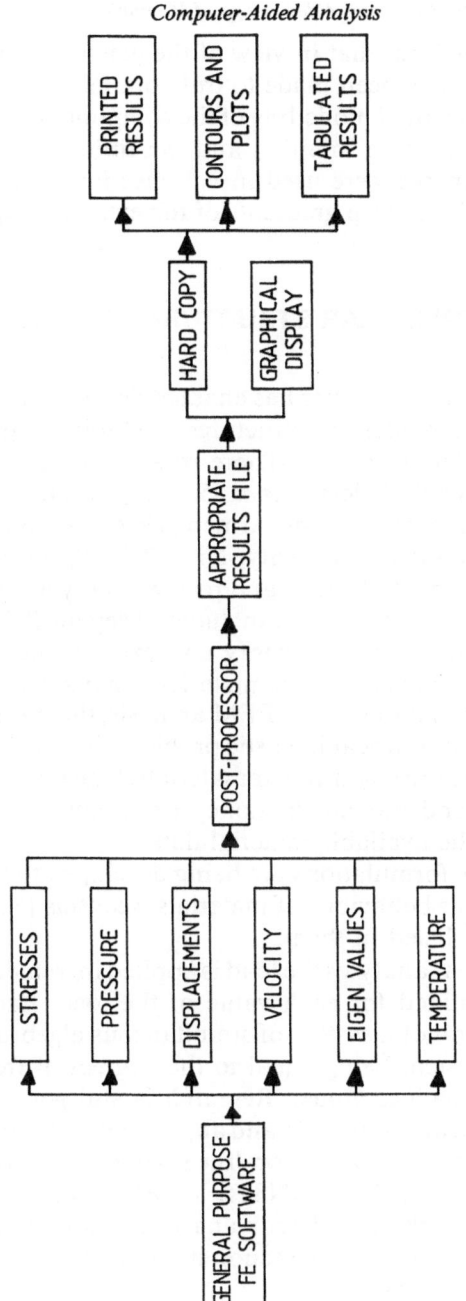

FIG. 5.2.—*contd.* (b) Post-processor interface.

It must be pointed out that in view of the extensive computer time needed, no attempt has been made to truly optimise the mesh system used in three-dimensional modelling of the components, other than the application of engineering intuition and the finite element expertise. High intensity elements were used in volumes involving geometrical discontinuities and for steep gradients of the externally applied loads.

5.5 NON-LINEAR STRUCTURAL ANALYSIS

The high speed digital computer has enabled designers and analysts to construct elaborate models of structures, including large deflection effects and material non-linearity. There are several excellent surveys of the various approaches: Tillerson *et al.* [5.8] review numerical methods used to solve non-linear equations; Armen [5.9] describes several analytical models of multiaxial plasticity; Nickell [5.10] gives a survey of techniques for treatment of creep, and reviews many widely used computer programs in which creep is included; Meguid [5.11] and [5.12] presents comparisons between theory and experiments for currently used models of elastic-viscoplastic material behaviour.

It is unfortunate that in current FEM analysis, the numerical ability to solve large-scale non-linear inelastic problems has advanced beyond the ability to characterise non-linear materials. One must constantly be vigilant not to spend too much money on analyses which are too sophisticated for the available material data.

New constitutive formulations are being developed to better characterise the non-linear behaviour of materials; see Refs [5.11] and [5.13] and the references listed in them.

In most non-linear analyses the load is applied incrementally and the response is determined for each value of the load. Each load level involves the solution of a system of simultaneous algebraic equations, the rank of this system being equal to the degrees of freedom in the discretised mathematical model. Research is still going on in many non-linear material analysis areas and appropriate solution algorithms. However, a number of reliable non-linear codes are currently in use; these include MSC/NASTRAN, ANSYS, MARC, ADINA and ABAQUS.

In the present work, only linear analysis was carried out on the different designs, and the above section has only been included for the purpose of completeness.

First Case Study: Computer-Aided Analysis
of Wheel-Assembly

5.6 STRUCTURAL STRESS ANALYSIS

In this section we only examine the critical components of the wheel-assembly. The original designs of impeller, discs and wheel hub, described in Chapter 3, were transferred to the analysis section of the package using the same data base. As a result of the detailed analysis the geometries of these components were modified. The following sections provide a detailed account of the results obtained for both the original and modified designs.

5.6.1 Geometry definition, model preparation and model checking
The solid modelling facility of SDRC (GEOMOD) enabled the three-dimensional definition of the examined models. Figure 5.3(a) shows

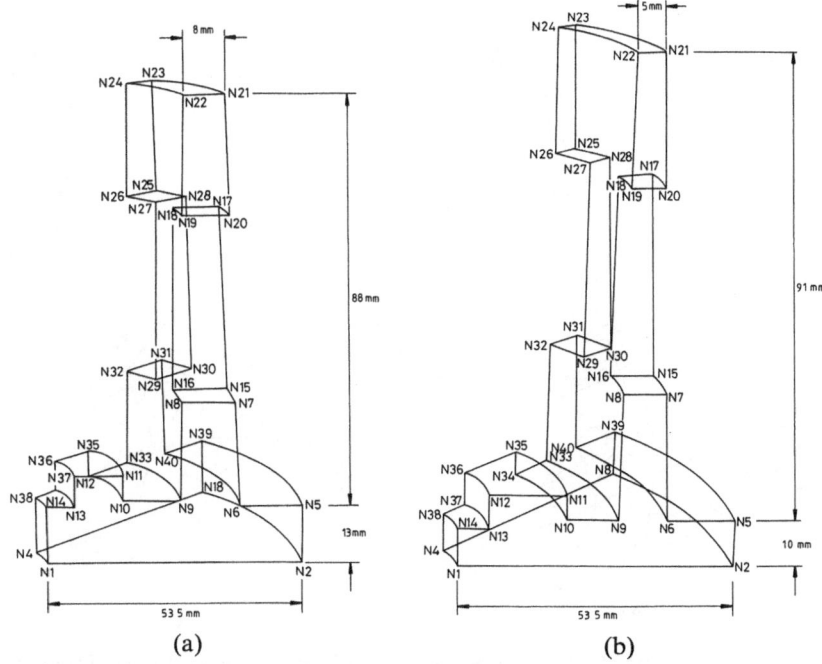

(a)   (b)

FIG. 5.3. 3-D solid model representation of a sector of peening impeller using SDRC-GEOMOD: (a) original design; (b) modified design.

a three-dimensional solid model representation of a sector of the originally proposed design of the impeller, while Fig. 5.3(b) shows the current modified geometry with the reduced dimensions. In view of the axisymmetric nature of the component, only one-eighth of its geometry needs to be considered for the analysis. These sectors were then transferred to the pre-processor (SUPERTAB) for meshing using the Enhanced Mesh Generation (EMG) facility provided in the SDRC package.

The two mesh systems used for the computation are shown in Fig. 5.4. For the original design, a total of 320 elements interconnected at 525 nodal points were used, and 563 elements with 803 nodal points were used for the modified design. In both cases, eight-noded solid brick elements were used.

A similar approach was adopted for the solid modelling and subsequent meshing of the wheel hub. In this case, the Mapped Mesh Generator (MMG) facility was used to mesh the models.

Figure 5.5(a) shows the solid model representation of a sector of the

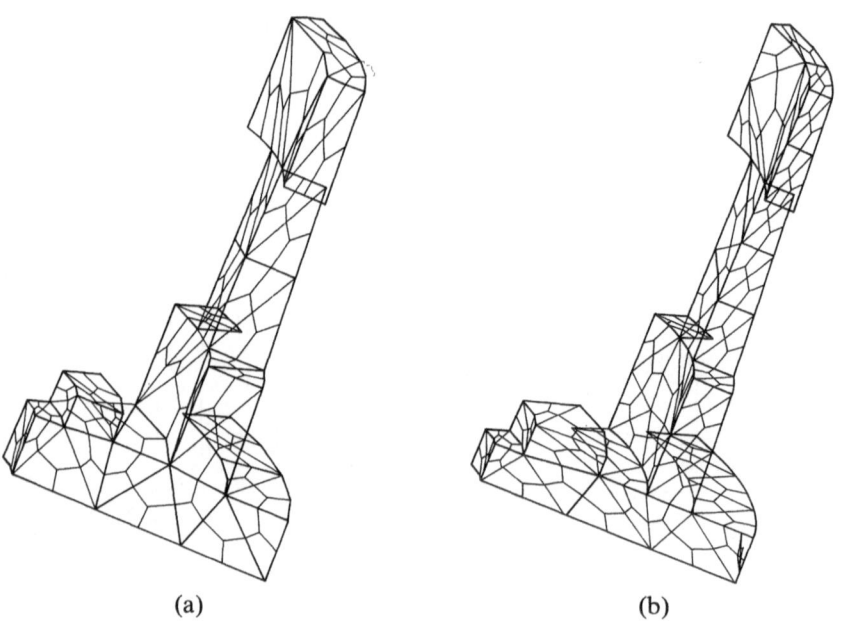

(a)   (b)

FIG. 5.4. Discretised structure of a sector of impeller using SDRC interactive 3-D Enhanced Mesh Generation (EMG) facility: (a) original design; (b) modified design.

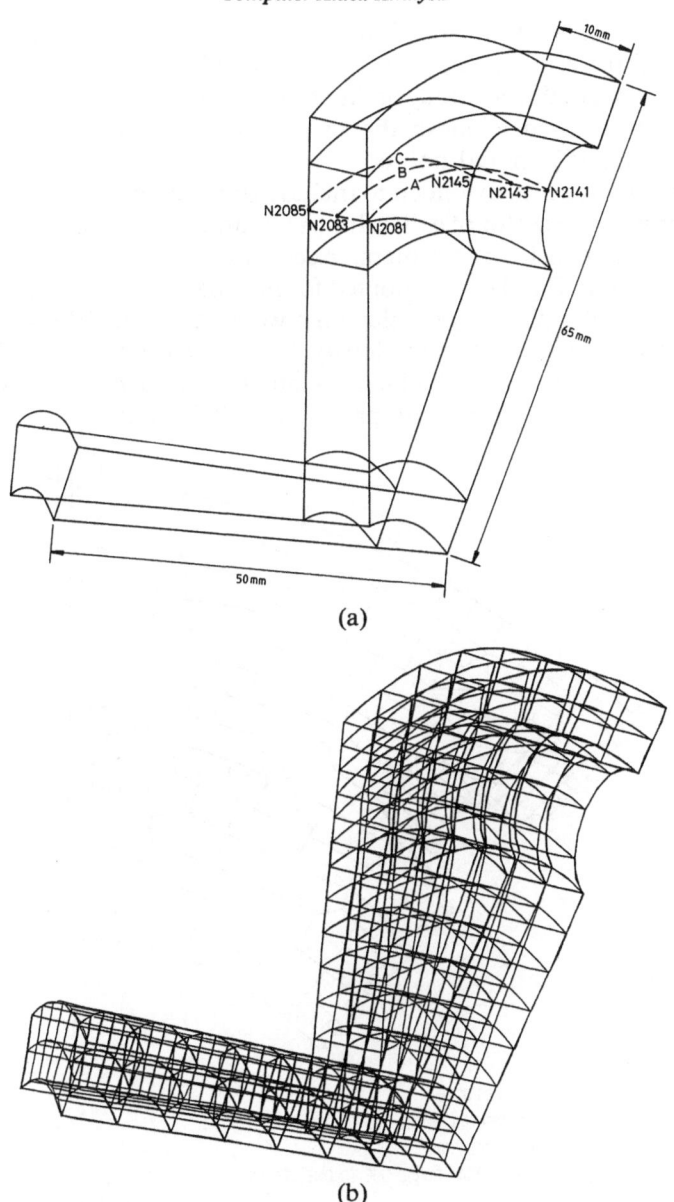

(a)

(b)

FIG. 5.5. (a) 3-D solid model representation of a sector of modified wheel hub; (b) discretised structure of a sector of modified wheel hub using SDRC interactive 3-D Mapped Mesh Generation facility (MMG).

modified design of the wheel hub, while Fig. 5.5(b) shows the corresponding three-dimensional discretised model of the component. In
order to improve the accuracy of the solution, 20-noded parabolic brick
elements were used to mesh the sector amounting to a total of 240
elements and 1533 nodal points.

The study was also extended to include finite element stress analysis
of an appropriate sector of a disc. It was felt unnecessary to treat the disc
as a three-dimensional problem. Accordingly, the two-dimensional
model shown in Fig. 5.6 was adopted for the analysis. In this case, eight-
noded parabolic plane stress elements were used with EMG.

Prior to the analysis, the previously discretised models were checked
for coincidence nodes, coincidence elements, element distortion and
free-edge continuity in the pre-processor SUPERTAB.

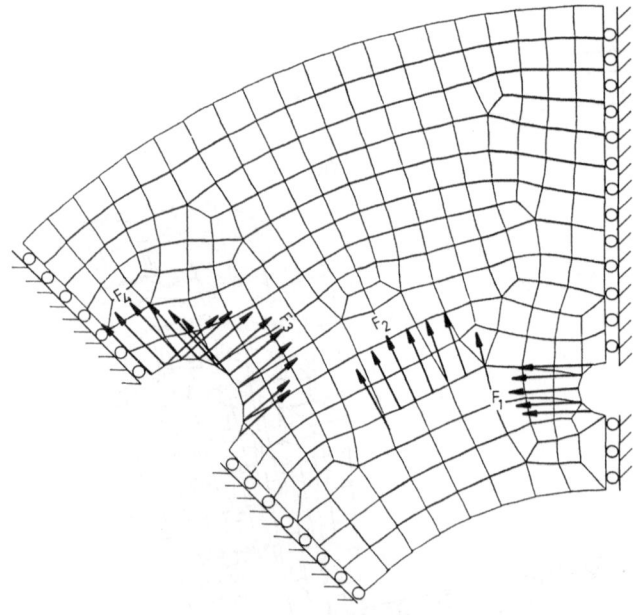

F_1	DRIVING FORCE
F_2	CENTRIFUGAL FORCE DUE TO ONE BLADE
F_3	RESISTING FORCE
F_4	CENTRIFUGAL FORCE DUE TO ONE SPACER

FIG. 5.6. Discretised structure of a sector of a rotating disc using 2-D Enhanced
Mesh Generation showing applied loads and imposed restraints.

5.6.2 Applied loads and imposed restraints

The following loads were considered during the structural stress analysis of the impeller:

(i) body forces applied as a result of a constant angular velocity of 3000 rev/min about the global z-axis. No allowances were given to account for the centrifugal forces resulting from the relatively small mass of shot accumulated in the impeller,

(ii) torque transmitted from the wheel hub to the impeller through the positioning groove (20 Nm); the resulting nodal forces were only a very small fraction of the body forces and were ignored in the analysis. Appropriate restraints in the tangential direction were imposed upon the discretised sector.

As regards the wheel hub, the following loads were assumed:

(i) body forces applied as a result of a constant angular velocity of 3000 rev/min about the global z-axis,

(ii) input torque of 144 Nm which is transmitted through the taper lock bush; the resulting shear stresses were insignificant in comparison with other applied loads,

(iii) torque transmitted at the bolts which amounts to 144 Nm and gives rise to a total force of 434 N,

(iv) bending moment resulting from the overhanging weight of the wheel. The magnitude of the resulting nodal forces was insignificant in comparison with (i) and (iii).

Again, the appropriate tangential restraints were applied interactively to the finite element model.

In the case of the disc, six loads were considered:

(i) body forces ($\omega_z = 3000$ rev/min);

(ii) input torque (144 Nm) which gives rise to $F_1 = 434$ N, as shown in Fig. 5.6;

(iii) centrifugal force due to the presence of a blade, $F_2 = 10$ kN;

(iv) resisting torque (144 Nm) which gives rise to $F_3 = 326$ N;

(v) centrifugal force due to the presence of a spacer, $F_4 = 6.3$ kN;

(vi) resisting force due to impinging shots. This was considered small in comparison with the previously mentioned loads.

Figure 5.6 shows the applied loads and the appropriate restraints imposed upon the examined sector.

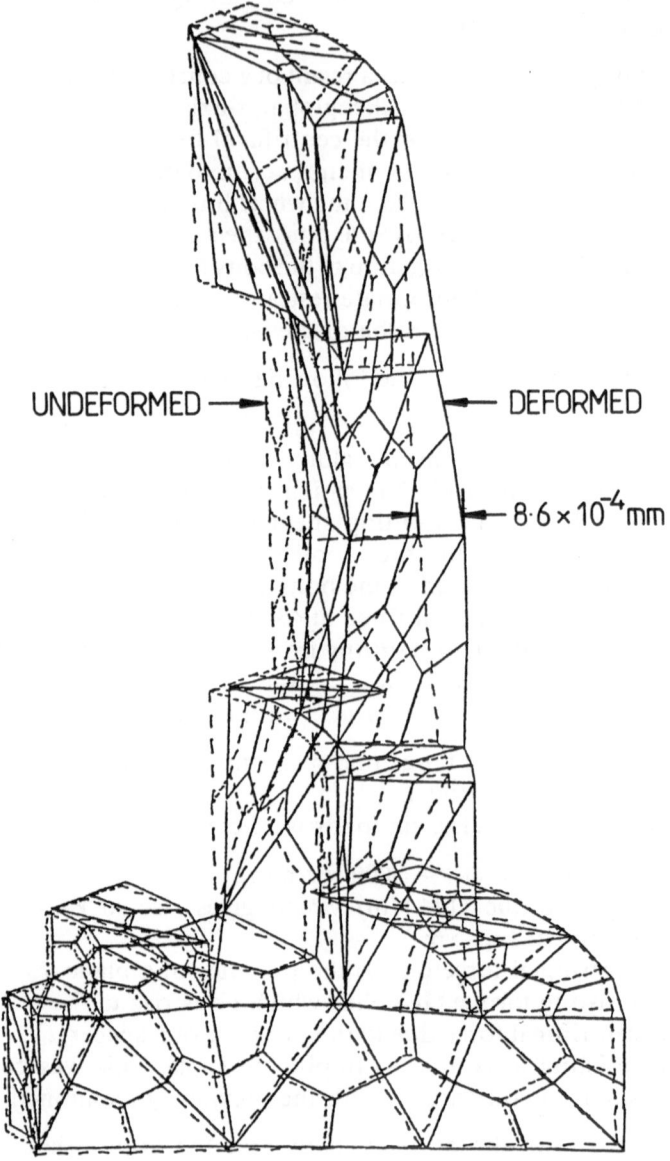

Fig. 5.7. Deformed shape of a sector of impeller: (a) original design.

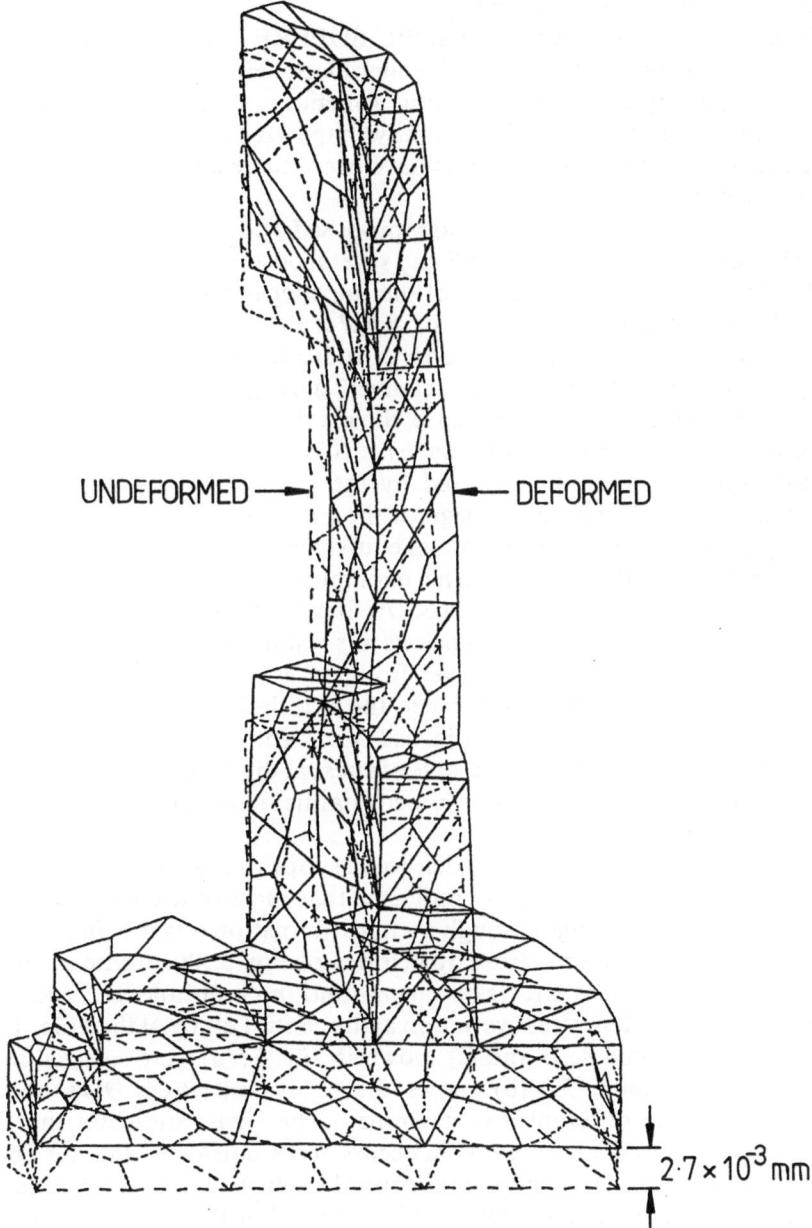

UNDEFORMED → ← DEFORMED

2.7×10^{-3} mm

FIG. 5.7.—*contd.* (b) Modified design.

5.6.3 Static stress analysis

The static analysis capability of SDRC-SUPERB is based on small displacement theory and assumes all materials are Hookean in behaviour. In this package, the interpolating functions of all isoparametric elements are of the 'serendipity' family shape functions as described in Ref. [5.14]. A number of attempts have been made and the outcome is described below.

Grid distortions of the originally proposed and modified designs of the impeller are shown in Fig. 5.7 and several observations can be made. The throat of the initial design of the impeller bulges outward with a maximum radial deformation of $8 \cdot 627 \times 10^{-4}$ mm, while the base experiences very little deformation as indicated by the solid lines of Fig. 5.7(a). Figure 5.7(b) shows a relatively more pronounced distortion in the modified design. In this case, the maximum deformation resulting from the assumed loads was $2 \cdot 7 \times 10^{-3}$ mm. The increase in the maximum deformation is caused by the proposed reduction in the dimensions of the original design of the impeller.

Figure 5.8(a) shows the von Mises equivalent stress $\bar{\sigma}$ contours occurring in the modified design. The maximum value of the equivalent stress occurs at the contours marked (A). It is worth pointing out that the proposed reduction in the dimensions has resulted in an increase in the maximum equivalent stress level from $3 \cdot 7$ MPa to $4 \cdot 6$ MPa. Figure 5.8(b) shows a comparison between the two designs for the different boundary nodes indicated in Fig. 5.3(a) and (b). Both diagrams show that there exists a non-uniform state of stress at the different positions of the model. In this case, however, a substantial saving in material was achieved ($\simeq 22\%$).

Similarly, Fig. 5.9 shows the deformed shape of the modified design of the wheel hub. This figure indicates that most of the deformation takes place in the flange through an inward rotation towards the collar section. The equivalent stress trajectories of Fig. 5.10(a) indicate that maximum equivalent stresses occur around the bolt-hole. In order to assess the equivalent stress variation at the central plane of the hole, the three arcs (marked A, B and C) shown in Fig. 5.5(a) were selected for comparison. Figure 5.10(b) shows the equivalent stress $\bar{\sigma}$ vs node number for the different arcs examined. The figure indicates that a maximum stress of about $4 \cdot 8$ MPa occurs at the outside surfaces A and C, while only $2 \cdot 5$ MPa occurs at the central arc B. As a result of the stress concentration, greater stresses will be induced at the top and bottom edges of the hole. The finite element results indicate that the maximum

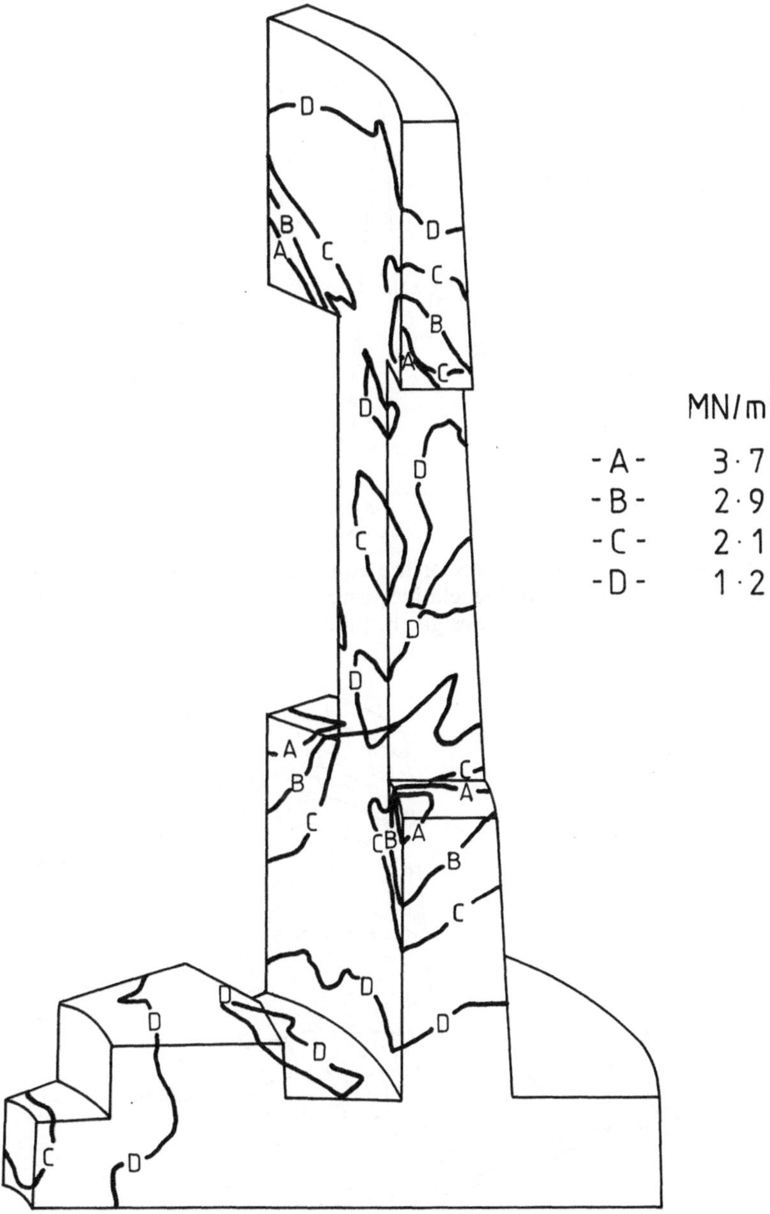

MN/m^2

- A -   3·7
- B -   2·9
- C -   2·1
- D -   1·2

FIG. 5.8. (a) Equivalent stress trajectories in the modified design of impeller showing regions of high stress concentration.

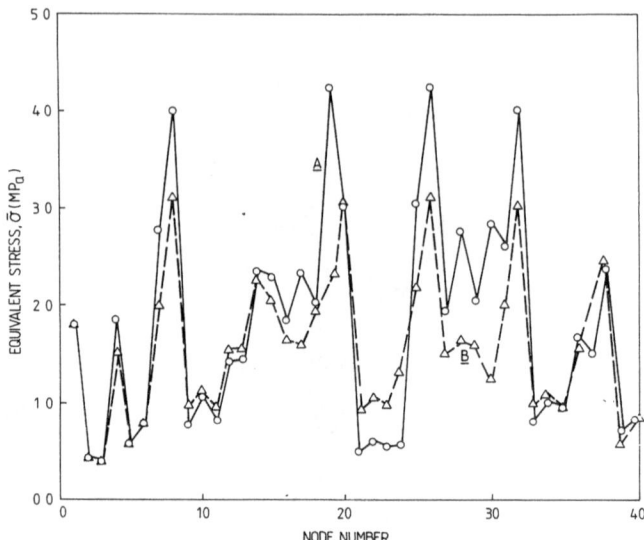

FIG. 5.8.—*contd.* (b) Plane plots of the equivalent stress vs node number in impeller (refer to Fig. 5.3 for the relative positioning of nodes in the structure): A, original design; B, modified design.

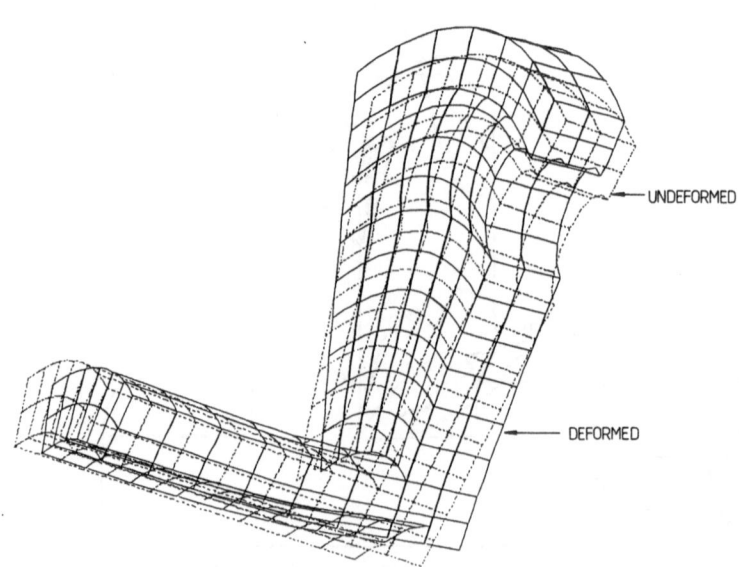

FIG. 5.9. Deformed shape of a sector of wheel hub.

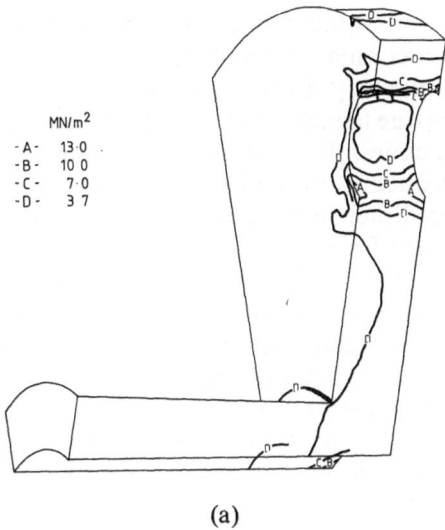

MN/m²
- A -   13·0
- B -   10·0
- C -   7·0
- D -   3·7

(a)

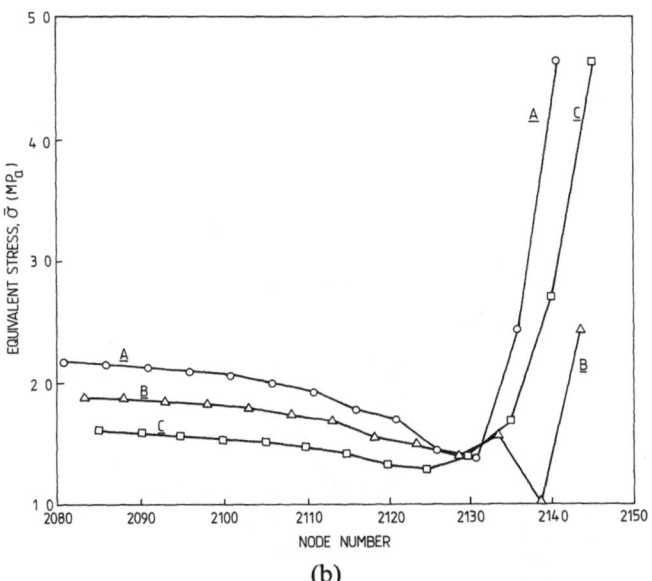

(b)

FIG. 5.10. (a) Equivalent stress trajectories in the modified design of the wheel hub showing regions of high stress concentration. (b) Plane plots of the equivalent stress vs node number in modified wheel hub (refer to Fig. 5.5(a) for the relative positioning of nodes in the structure).

equivalent stress level at the bottom of the hole in the modified wheel hub is 16·7 MPa in comparison with 29·15 MPa in the originally proposed design. It must be pointed out that in this case the reduction in the stress level was due to reduction in body forces caused by the reduction in flange thickness from 20 mm to 10 mm. As a result of this modification, a reduction in total weight of 40% was achieved.

Finally, the two-dimensional analysis of the disc indicates that the maximum $\bar{\sigma}$ increases from 59 MPa to 75 MPa as a result of a reduction in its thickness from 16 mm to 12 mm. Figure 5.11 shows the highly stressed regions in the disc. Again, the proposed reduction in thickness induces a substantial reduction in disc weight (amounting to 24%) without sacrificing its structural integrity.

Table 5.2 summarises the results of the critical components examined and the relevant savings as well as maximum stress levels and maximum deformation achieved as a result of the proposed modifications to the different designs.

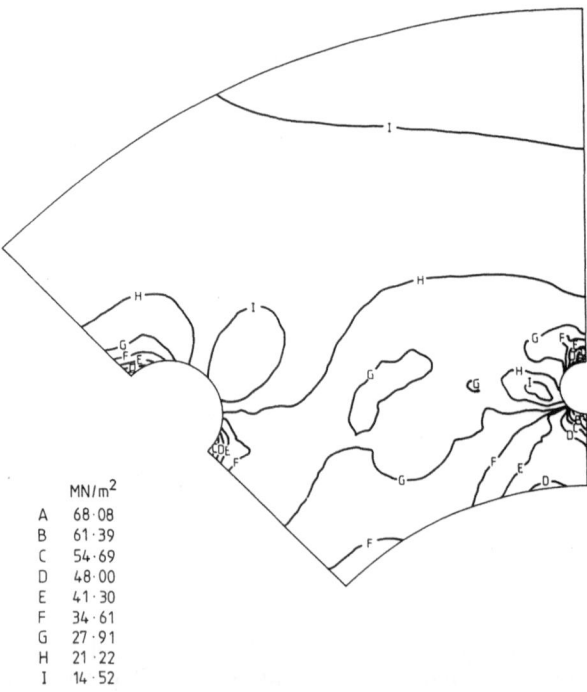

	MN/m²
A	68·08
B	61·39
C	54·69
D	48·00
E	41·30
F	34·61
G	27·91
H	21·22
I	14·52

FIG. 5.11. Equivalent stress trajectories in disc showing regions of high stress concentration.

Table 5.2
Summary of Results of Critical Components

Component Item	Impeller		Wheel hub		Disc	
	Original	Modified	Original	Modified	Original	Modified
Weight (kgf)	2·6	2·02	4·55	2·68	6·52	4·59
Maximum equivalent stress, $\bar{\sigma}$ (MPa)	3·42	4·56	29·15	16·70	58·67	74·77
Maximum deformation (mm)	$8·6 \times 10^{-4}$	$2·7 \times 10^{-3}$	$1·68 \times 10^{-2}$	$1·69 \times 10^{-2}$	$1·0 \times 10^{-2}$	$1·26 \times 10^{-2}$

5.7 DYNAMIC RESPONSE STUDIES

5.7.1 Modelling of the assembly

In addition to the conventional approach of Rayleigh, the finite element method was adopted in the determination of the first bending mode of the simplified wheel-assembly shown in Fig. 5.12. In the FE computations, the assembly, which is shown schematically in Fig. 5.13(a), was reduced to 34 linear beam elements with appropriate lumped mass and supporting spring elements, as shown in Fig. 5.13(b). The corresponding section properties are given in Table 5.3.

Table 5.3
Cross-Sectional Properties of the Wheel-Assembly

Element	Physical property	Area $(m^2 \times 10^{-5})$	I_y $(m^4 \times 10^{-9})$	I_z $(m^4 \times 10^{-9})$
1		196	307	307
2		283	636	636
3		332	876	876
4		180	261	261
5		554	2444	2444
6		2898	410	410
7		6038	1159	1159
8		385	469 (46860)	469 (46860)

The choice of bearings was dictated by the imposed static and dynamic loads as well as the other relevant design constraints. The radial and axial stiffnesses of the proposed deep groove rolling element bearings (SKF designation number 6410) were obtained from a software developed by SKF [5.15]. The following values were used in the analysis: for the radial stiffness $K_r = 1 \cdot 8 \times 10^8$ N/m and for the axial stiffness $K_a = 9 \cdot 2 \times 10^6$ N/m. In the above calculations, no allowances were given to account for possible gyroscopic effects since they will have a stiffening effect on the shaft and accordingly increase its fundamental frequency. Torsional vibrations were not attempted using FEM. However, simplified calculations reveal that no torsional vibration problems will be encountered with the present design.

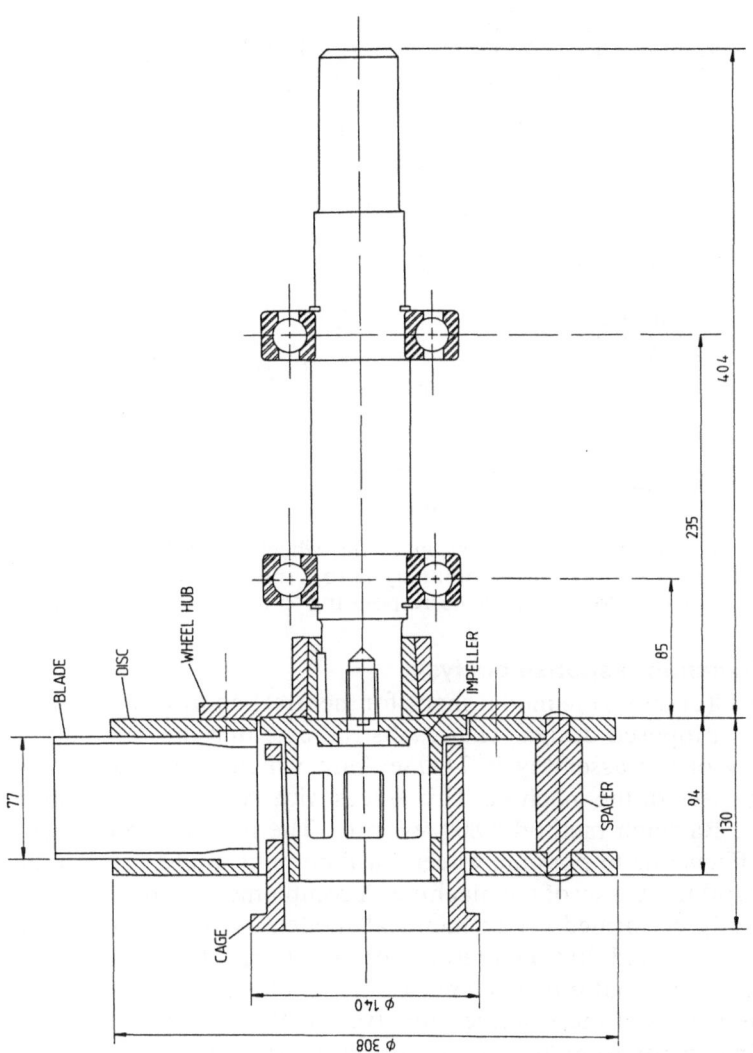

Fig. 5.12. A computer representation of 2-D engineering drawing of the wheel-assembly used in the dynamic response studies (all dimensions in mm).

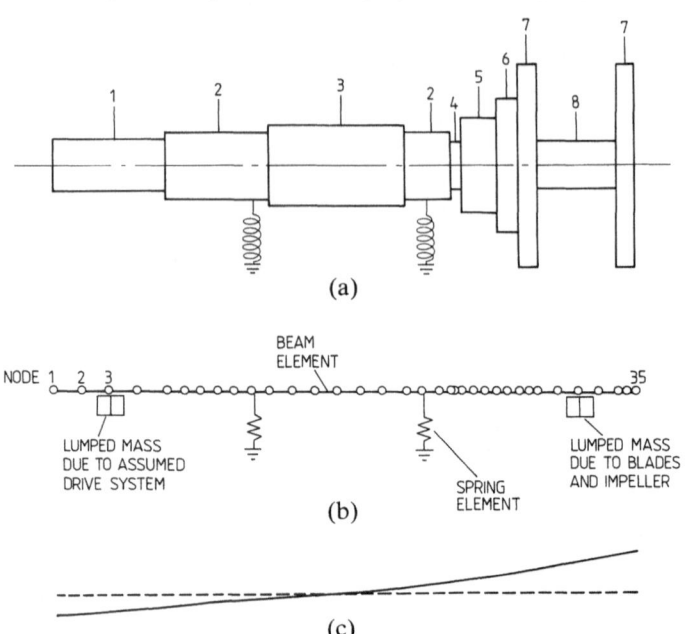

FIG. 5.13. (a) Schematic representation of the wheel-assembly showing the different sections used in the dynamic analysis. (b) Finite element discretisation of the assembly. (c) Modal shape corresponding to the first bending mode.

5.7.2 Dynamic response analysis

Figure 5.13(c) shows the modal shape for the first bending mode for the originally proposed design. The results indicate that the first natural frequency of the assembly is 7440 rev/min which is well above the running speed of the machine. This agrees with the 'hand-calculation' using the Rayleigh method (9800 rev/min). The introduction of gyroscopic effects due to possible wheel misalignment would increase the first natural frequency of the machine. A complementary dynamic response study performed on an identical model, using transfer matrix technique, revealed that the presence of gyroscopic effects will indeed increase the first natural frequency of the machine by approximately 15%. The details of these studies are given in Refs [5.16] and [5.17].

It is interesting to note that the work was extended to examine the effect of varying the stiffness of the overhang upon the dynamic response of the system. The results of Table 5.4 reveal that the effect of such variation upon the natural frequencies is insignificant.

The proposed modified geometries of the wheel-assembly were also subjected to similar dynamic response studies. The results indicate that no vibration problems will be experienced (ω_{n1} = 8040 rev/min) either at or near the running speed of the machine which is 3000 rev/min.

Table 5.4

List of Natural Frequencies for Two Different Values of Second Moment of Area for the Overhang

Number	Second moment of area about y and z axes I_y and $I_z \times 10^{-9} \, m^4$	
	First assumed value of I for the overhang (46860)	Second assumed value of I for the overhang (469)
	Frequency (Hz)	*Frequency (Hz)*
1	0·12389E+03	0·12508E+03
2	0·25601E+03	0·25641E+03
3	0·79804E+03	0·88168E+03
4	0·17460E+04	0·17125E+04
5	0·38045E+04	0·32706E+04
6	0·61193E+04	0·41752E+04
7	0·65799E+04	0·62641E+04
8	0·71548E+04	0·82373E+04
9	0·83665E+04	0·11098E+05
10	0·11685E+05	0·13727E+05
11	0·15543E+05	0·17085E+05
12	0·18734E+05	0·19741E+05
13	0·22194E+05	0·21570E+05
14	0·26836E+05	0·26480E+05
15	0·30410E+05	0·30406E+05

Second Case Study: Computer-Aided Analysis of Fluid Coupling

5.8 STRUCTURAL STRESS ANALYSIS AND DYNAMIC RESPONSE STUDIES

Due to the complex nature of the critical components which constitute the coupling assembly, it was necessary to undertake a finite element investigation. The critical components of the preliminary design described earlier in Fig. 3.22 of Chapter 3 were transferred from the object modelling section of **SDRC-GEOMOD** to the finite element preprocessor for model creation, preparation and checking.

A similar approach to that adopted in the meshing of the critical components of the wheel-assembly was followed, and Figs 5.14, 5.15 and 5.16 show the discretised three-dimensional models of the coupling: casing, impeller and rotor, respectively. In anticipation of the high stress gradients at the casing hub, a finer mesh was generated in that area. High stresses were also expected where the blade joins the torus and the shroud of both impeller and rotor. Again, a finer mesh was used for the torus but the geometry dictates a coarser mesh for the shroud. The above discretised models were checked for coincidence nodes, coincidence elements, element distortion and free-edge continuity in the preprocessor. In all the above cases, the element used for the three-dimensional analysis was the isoparametric 20-noded parabolic solid element. This element allows for the accurate representation of the model boundaries and yields results of sufficient accuracy at a reasonable expenditure of computer resources.

The complex nature of the pressure gradient acting within the coupling requires certain simplifying assumptions to be made. These include

  (i)   the working fluid is assumed incompressible and the cavitation phenomenon does not occur within the closed circuit,
  (ii) steady state conditions dominate the flow within the coupling,
  (iii) the pressure distribution over the blades is assumed uniform,
  (iv) the fluid angle is always equal to the blade angle at entrance and exit from each element,
  (v)   the static head due to the weight of the fluid is negligible.

The following loads were then considered during the stress analysis of the different critical components:

  (i)   the supply pressure (280 kN/m^2),
  (ii) the pressure due to the forced-vortex about the coupling's rotational axis (which is a function of r^2), which varies from 52 kN/m^2 at the outer radius of the shaft to 950 kN/m^2 at R_4 (see Fig. 3.19),
  (iii) the forced-vortex about the toroid axis (2 kN/m^2),
  (iv) the dynamic pressure on the blades causing rotation (90 kN/m^2) — see Ref. [5.18],

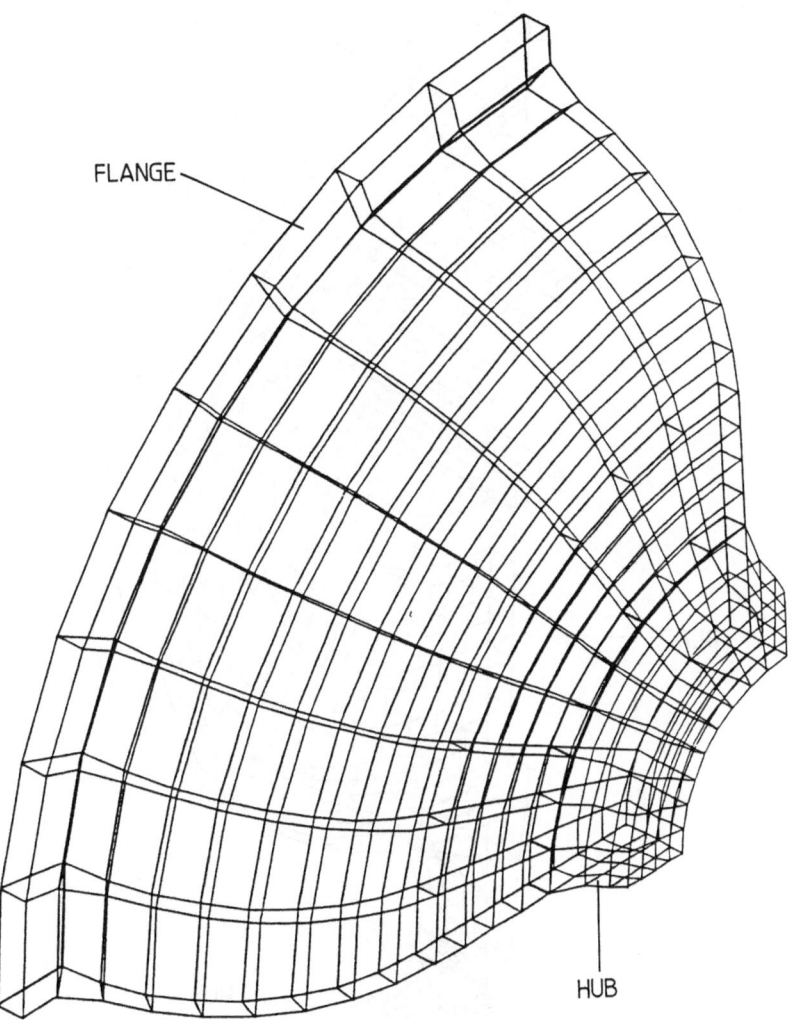

FIG. 5.14. Discretised structure of casing using three-dimensional parabolic brick elements.

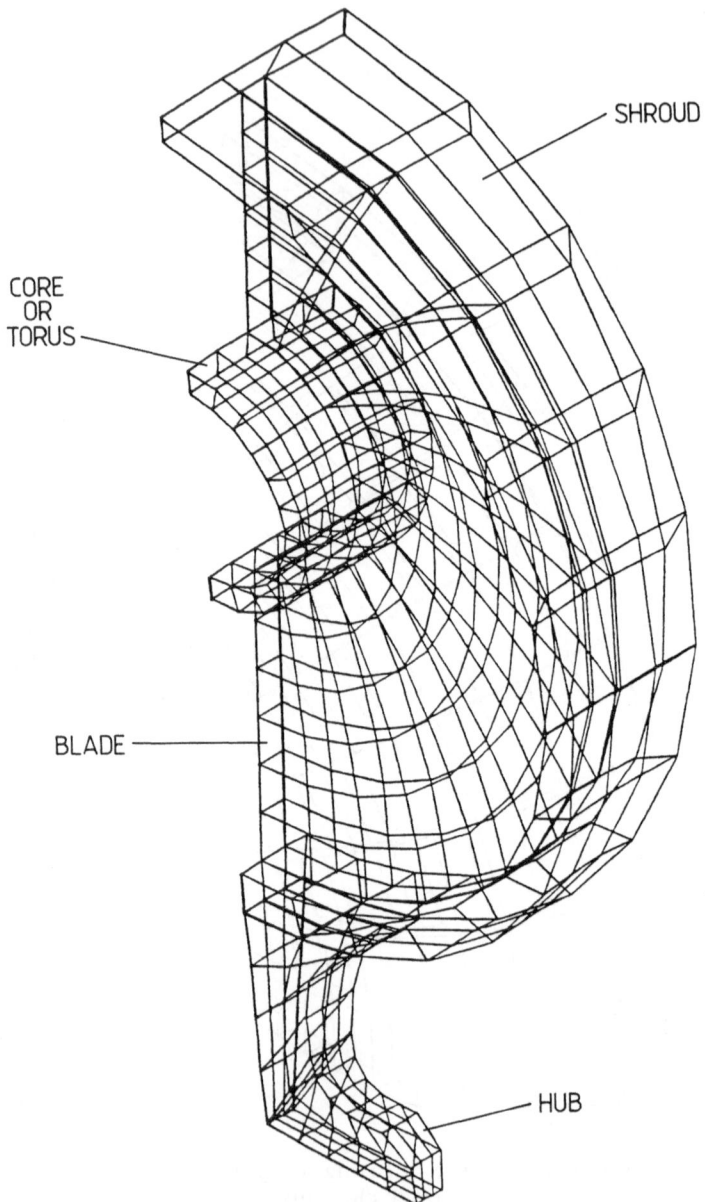

CORE
OR
TORUS

SHROUD

BLADE

HUB

FIG. 5.15. Discretised structure of impeller using three-dimensional parabolic brick elements.

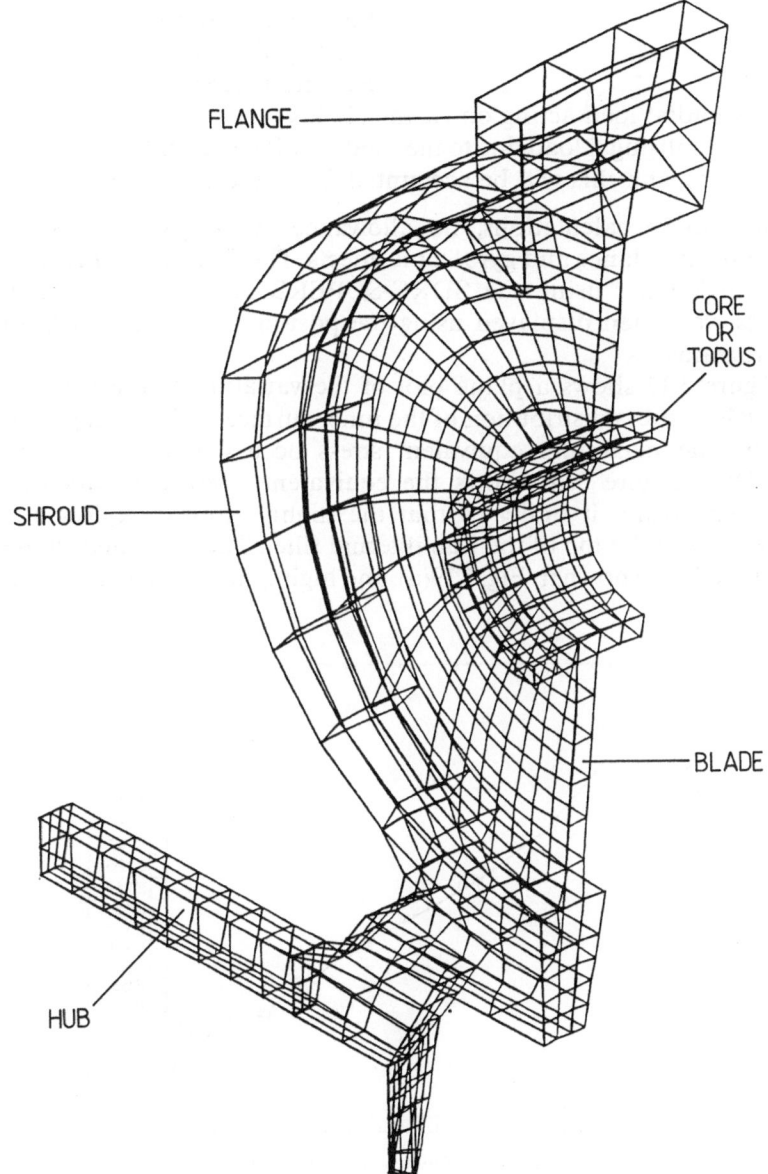

FLANGE

CORE
OR
TORUS

SHROUD

BLADE

HUB

FIG. 5.16. Discretised structure of rotor using three-dimensional parabolic brick elements.

(v)   reactions at the bearings = Bearing A (628 N)

Bearing B (697 N)

Bearing C (314 N)

Bearing D (383 N)

(vi)   radial load acting on casing (20 kN),

(vii)   centrifugal load due to the rotation of the coupling (function of r^2), and this will be accounted for in the FE analysis.

In the above calculations, no allowances were given to possible dynamic unbalance and gyroscopic effects. The finite element analysis was carried out using the SDRC-SUPERB analysis program. The results were then displayed using the post-processor of SDRC suite of programs.

Figure 5.17 shows a plane plot of the variation of the von Mises equivalent stress distribution of the inner surface of the casing. In this region, the maximum equivalent stress occurs at the hub section (33 MPa). Figure 5.18 shows the equivalent stress contours in the impeller. Again, it indicates that the highly stressed region of the impeller is at the top of the hub internal fillet. The deformed shape of the impeller is shown in Fig. 5.19(a), and highlights the fact that bending

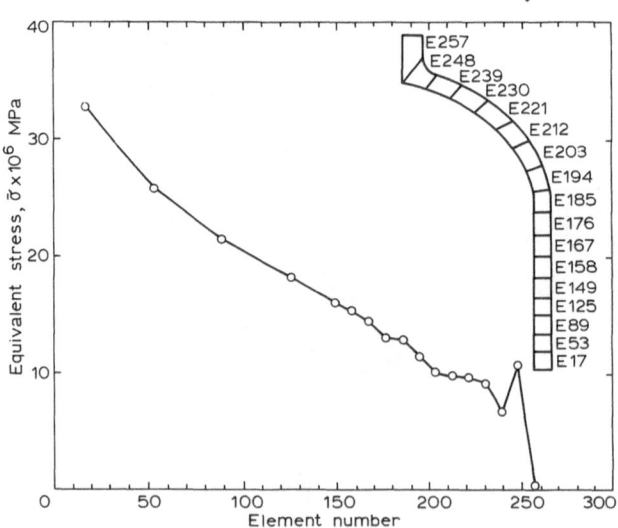

FIG. 5.17. A plane plot of the von Mises equivalent stress variation from the hub to the flange of the casing.

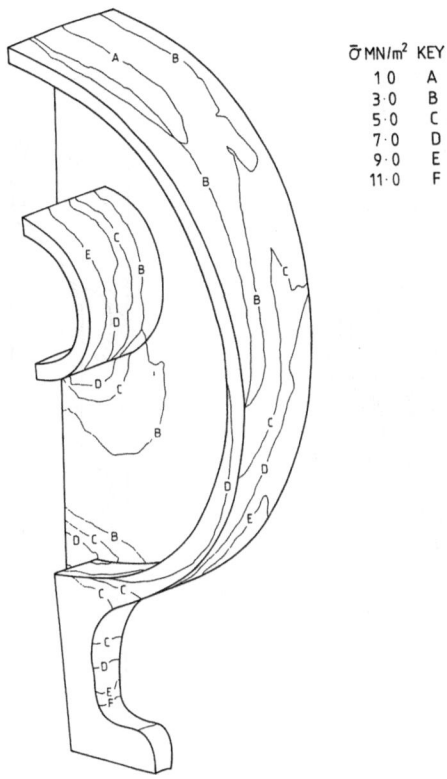

σ̄ MN/m²	KEY
1·0	A
3·0	B
5·0	C
7·0	D
9·0	E
11·0	F

FIG. 5.18. Equivalent stress contours in the impeller.

effects will take place at the indicated section (A–A). It also indicates that the maximum deformation near the bearing is within the maximum allowable deformation. However, it was felt that the addition of stiffening webs, near the hub, to both impeller and rotor would enhance their strength and possibly their dynamic characteristics. The proposed modified design of the impeller, with added stiffening webs, is shown in Fig. 5.19(b).

The study has also been extended to examine the dynamic behaviour of the coupling. In order to perform the analysis, it was necessary to model the coupling as spring, mass and beam elements, as shown in Fig. 5.20(a).

The results of the dynamic analysis indicate that the first natural

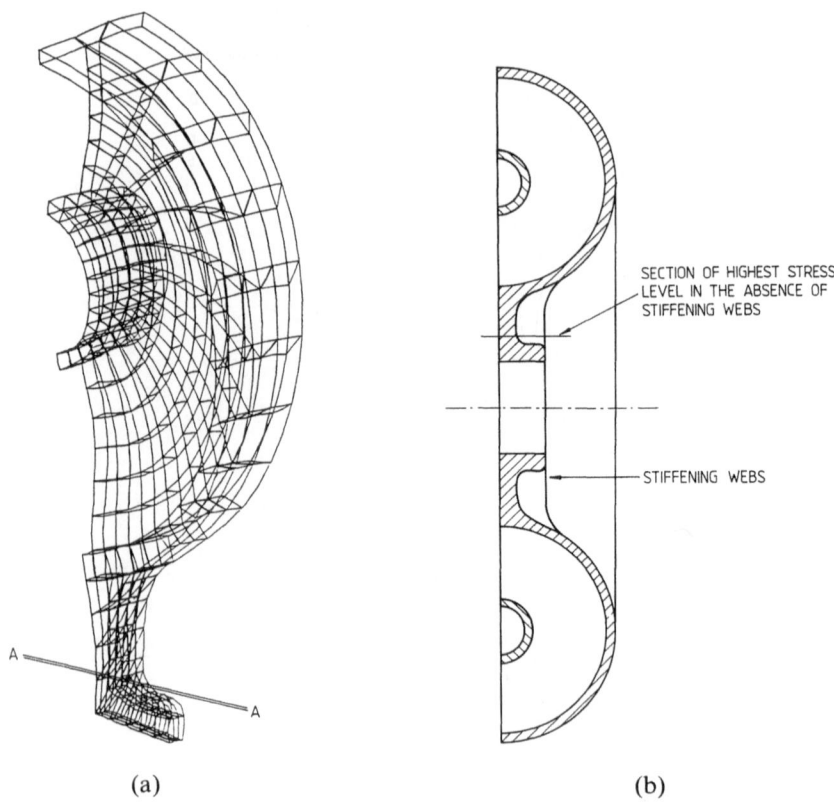

(a)   (b)

FIG. 5.19. (a) The deformed shape of the impeller (maximum displacement = 0·212 mm). (b) Proposed modified design of impeller.

frequency existed at 6420 rev/min. The accompanying bending mode shape is shown in Fig. 5.20(b). It is also desirable, when designing rotating machinery, to avoid operating the machine at 30% below or 20% above the first bending mode of vibration. It appears from the present calculations that, with an input running speed of 1500 rev/min, no vibration problems will be experienced with the present design. It is also worth pointing out that it is good design practice to avoid natural frequencies 10% below and 15% above twice the running speed of the machine. Inspection of the second natural frequency indicates that the above criterion has also been satisfied with the present design.

ROTOR AND CASING

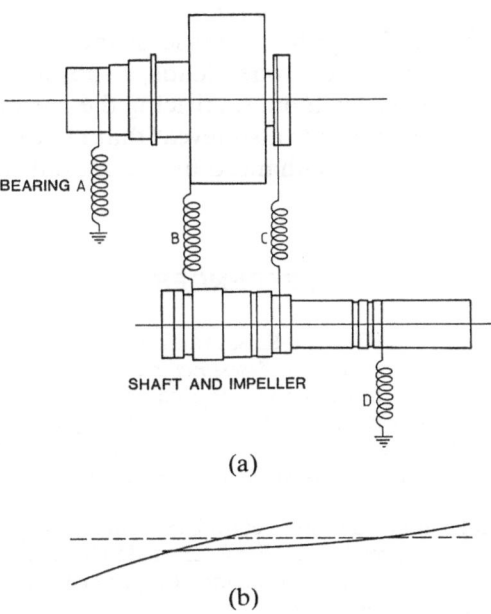

(a)

(b)

FIG. 5.20. (a) Schematic layout of the coupling for the dynamic analysis. (b) Predicted mode shape.

5.9 CONCLUSIONS

The above case studies demonstrate the efficient utilisation of computer-aided three-dimensional finite element analysis in the design of the critical components of (i) a wheel-assembly for a centrifugal peening equipment and (ii) a constant-fill fluid coupling. Throughout the investigation, the appropriate geometries were generated using the interactive solid modelling facility of SDRC-GEOMOD. The three-dimensional mesh definition, model checking, model preparation and output display were all performed using SDRC pre- and post-processors of SUPERTAB. Both structural stress analysis and dynamic response studies were carried out using a general purpose finite element package known as SUPERB.

The study also reveals that the different aspects of model definition, analysis and output display represent a continuum of activity rather than a collection of separate functions; this was clearly emphasised by the use of one common data base.

As a result of examining the response of the critical components of the wheel-assembly to the applied loads, a substantial reduction in weight has been achieved without sacrificing the mechanical integrity of both assemblies. The results also reveal that the dynamic response of the assemblies would be enhanced by the introduction of the new modified geometries.

REFERENCES

[5.1]   W. Pilkey, H. Schaeffer and K. Saczalski (Editors), *Structural Mechanics Computer Programs: Surveys, Assessments and Availability*, University of Virginia Press, Charlottesville, Virginia, 1974.

[5.2]   H. G. Schaeffer, A review of the International Symposium on Structural Mechanics Software, *Computers and Structures*, **8**, pp. 589–98, 1978.

[5.3]   S. S. Rao, *The Finite Element Method in Engineering*, Pergamon Press, Oxford, 1982.

[5.4]   H. H. Fong, An evaluation of eight U.S. general purpose finite element computer programs, *Proc. 23rd AIAA/ASME/ASCE/AHS Structures, Structural Dynamics and Materials Conference, Part 1*, 10–12 May, Louisianna, pp. 145–60, 1982.

[5.5]   B. Fredricksson, J. Mackerle and B. G. A. Persson, Finite element programs in integrated software for structural mechanics and CAD, *Computer-Aided Design*, **13**, pp. 27–39, 1981.

[5.6]   R. E. Nickell, The Interagency Software Evaluation Group: A Critical Evaluation of the ADINA, NASTRAN and STAGS Structural Mechanics Computer Programs, Report to the Office of Naval Research, Arlington, Virginia, Contract No. 0014-79-C-0620, 1981.

[5.7]   Enhanced Mesh Generation, Model Solution, Data Loader and Output Display Manual, Structural Dynamics Research Corporation, General Electric, CAE International, USA, 1983.

[5.8]   J. R. Tillerson, J. A. Stricklin and W. E. Haisler, Numerical methods for the solution of non-linear problems in structural analysis, *Numerical Solution of Non-linear Structural Problems*, edited by R. Hartung, *AMD*, **6**, ASME, NY, pp. 67–101, 1973.

[5.9]   H. Armen, Plastic analysis, *Structural Mechanics Computer Programs*, edited by W. Pilkey, H. Schaeffer and K. Saczalski, University of Virginia Press, pp. 37–79, 1974.

[5.10] R. E. Nickell, Thermal stress and creep, *Structural Mechanics Computer Programs, Ibid.*, pp. 103–22, 1974.

[5.11] S. A. Meguid and L. E. Malvern, An experimental investigation into the plastic flow and strain-hardening of mild steel under proportional and abruptly changing deformation paths at a controlled rate, *ASME, J. Engng Mater. Technol.*, **105**(3), 147–54, 1983.

[5.12] S. A. Meguid, Plastic flow of mild steel (En8) at different strain-rates under abruptly-changing deformation paths, *J. Mech. Phys. Solids,* **29**, 375–95, 1981.

[5.13] J. A. Stricklin and K. J. Saczalski (Editors), Constitutive equations in viscoplasticity: computational and engineering aspects, *AMD,* **20**, ASME, NY, 1976.

[5.14] V. T. Nicholas and E. Citipitiogul, A general isoparametric finite element SDRC SUPERB, *Computers and Structures,* **7**, pp. 303–23, 1977.

[5.15] N. G. Stimson, Chief Application Engineer, SKF (UK) Ltd, private communications.

[5.16] M. S. Klair, Computer-Aided Engineering of Centrifugal Peening Equipment and Effect of Incomplete Coverage upon Component Fatigue Performance, PhD Thesis, Cranfield Institute of Technology, 1985.

[5.17] P. F. Abrahams, Computer-Aided Engineering of Centrifugal Peening Equipment, MSc Thesis, Cranfield Institute of Technology, 1985.

[5.18] W. May, Computer-Aided Analysis of a Fluid Coupling for an Auxiliary Feed Pump Application, MSc Thesis, Cranfield Institute of Technology, 1984.

Chapter 6

Computer-Integrated Manufacturing

In an integrated CAD/CAM system, a central unified data base is used to hold the design, analysis and manufacturing information. This means that the information is consistent throughout the Company. Once the product has been designed and analysed, much manufacturing information can be generated automatically.

In this section of the work, we plan to develop the data base needed for the manufacture of some of the critical components of the wheel-assembly and the fluid coupling. Specifically, emphasis will be given to the use of interactive graphics in the development of numerically controlled part-programming and the integration between the design and manufacturing activities. However, before discussing these important aspects, let us provide a brief account of the use of computers in modern manufacturing systems; a detailed description of this aspect is provided in Refs [6.1] to [6.5] and the references listed in them.

6.1 COMPUTER-CONTROLLED MACHINE TOOLS

In the past, engineering companies manufactured their products with manually operated machine tools and a large labour force. These manually operated machine tools were characterised by their relative simplicity and high level of reliability.

The early numerically controlled (NC) machines provided considerable increase in productivity compared with manual ones and, as such, could be justified on the basis of reducing labour costs. A similar situation occurred with early first generation computers where they were mainly used for costing, stock control and other clerical functions, and where the justification was mainly to eliminate highly labour intensive clerical activities.

The revolutionary change in component/system design, production techniques, maintenance, development and management and competitiveness which is predicted to take place by the end of this century will require unprecedented involvement of computer-controlled systems in engineering. Every activity in this factory of the future, including design, analysis, manufacture, assembly and inspection will be monitored and controlled by computers; some of these duties will be performed by robots and supported by intelligent systems, see Refs [6.1] and [6.2]. This will undoubtedly lead to improved quality of products, increased flexibility of production, reduced scrap and re-work and improved response to customers' needs.

Modern manufacturing systems and industrial robots are advanced automation systems which utilise computers as an integral part of their control. Computers are now a vital part of their control. They can control individual manufacturing systems, run integrated production lines and are beginning to take over control of an entire factory, as detailed in Refs [6.3] to [6.5].

The new era of automation, which started with the introduction of numerically controlled (NC) machine tools, was undoubtedly stimulated by computers. As a matter of fact, computers enabled the design of more flexible automated systems; namely, systems which can be adapted by programming to effectively produce a new product in a short time. Actually, flexibility is the key word which characterises the new era in automation of manufacturing systems.

Generally, manufacturing systems can be divided into small individual equipment, e.g. computer numerically controlled (CNC) machine tools, and comprehensive systems with appropriate manufacturing cells known as flexible manufacturing systems (FMS) and containing many individual systems. Both types of systems are controlled either by a computer or a controller based on digital technology. They can accept data in the form of programs and are able to process it and provide command signals to the prime movers and actuators which drive slides, rotary axes or material-handling systems. In 'stand-alone' individual systems, the input data defines the position of moving slides, velocities, type of motion, etc. In more sophisticated manufacturing cells, the system makes decisions based upon the feedback signals from the appropriate transducers to the associated robot ([6.4] and [6.5]).

Controlling a machine tool by means of a 'prepared' program is known as numerical control (NC). In a typical NC system, the numerical data which is required for producing a part is maintained on a punched tape and is called part-program. The part-program is arranged in the form of blocks of information, where each block contains the numerical data required to produce one segment of the component. These data include coded information concerning segment geometry, cutting speeds, feeds and depth of cut. Compared with a conventional machine tool, an NC system replaces the manual actions of the operator.

Initially, the preparation of the part-program for NC machine tools requires a part-programmer. Ideally, this part-programmer must possess the appropriate knowledge and experience in the mechanical engineering field, including knowledge of machine tools, cutting fluids, fixture design techniques, use of machinability data, and optimal sequence of operation. The part-program is written manually or by using a computer-assisted language, such as Automatically Programmed Tools (APT) [6.3].

In NC machine tools each axis of motion is equipped with a separate driving device which replaces the hand-wheel of the conventional machine. The driving device may be a dc motor, a hydraulic actuator or a stepping motor. The NC machine tool system contains the machine controller unit and the machine tool itself, as shown in Fig. 6.1. The machine controller unit consists of the electronics and hardware which read and decode the part-program into mechanical actions of the machine tool. The typical elements of a conventional NC controller unit include the tape reader, a data buffer, signal output channels to the machine tool, feedback channels and the sequence controls to co-ordinate the overall operation of the foregoing items. It should be noted that nearly all modern NC systems today are sold with a micro-computer as the controller unit. This type of NC is called computer numerically controlled (CNC) machine tools.

6.2 TYPES OF CONTROL SYSTEMS FOR NC MACHINES

There are two principal types of control systems for NC machines; the first is the point-to-point (PTP) positioning system, sometimes referred to as numerical positioning control (NPC), and the second is the numerical contouring control (NCC) system [6.3].

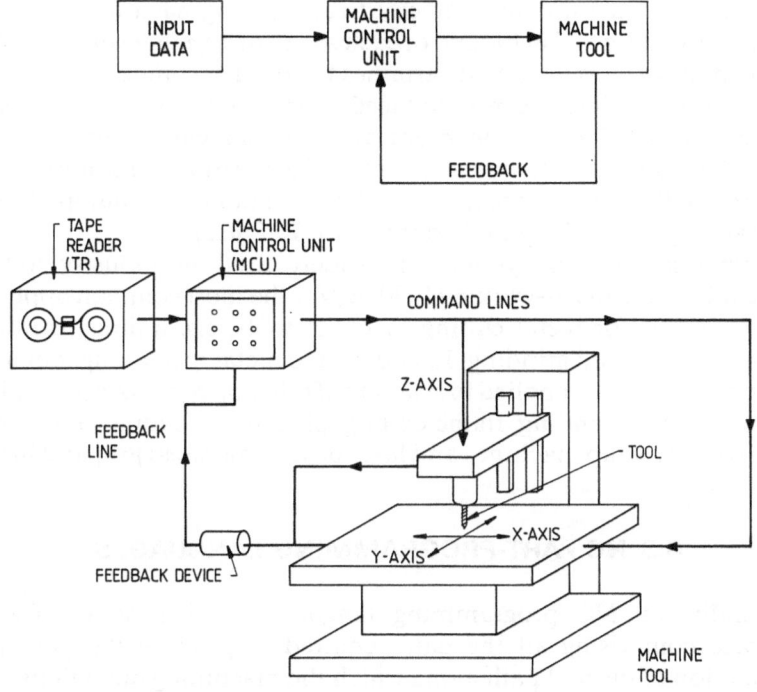

FIG. 6.1. Basic components of a conventional numerical control system.

The function of NPC systems is to move the machine table or spindle to a specified position so that machine operations may be performed at that point. In this case, the path taken to arrive at this point is unimportant, except to avoid collision, since no cutting is performed until after the point of interest has been reached. Therefore, this system would require position counters only for controlling the final position of the tool upon reaching the point of interest.

An improvement to the PTP system is the straight cut NC system in which the cutting tool moves parallel to the major axes at a controlled rate suitable for machining. It is therefore suitable for performing milling operations to fabricate workpieces of rectangular configuration. With this type of NC system it is not possible to combine movements in more than a single-axis direction. Therefore, angular cuts on the workpiece would not be possible.

In contouring or continuous path systems, the tool is cutting while the axes of motion are moving. All axes of motion might move simultaneously, each at a different velocity. In contouring machines, the position of the cutting tool at the end of each segment, together with the ratio between the axial velocities, determines the desired contour of the part, and at the same time, the resultant feed affects the surface finish. Since, in this case, a velocity error in one axis causes a cutter path position error, the system has to contain continuous-position control loops in addition to the position counters. Each axis of motion is equipped with a separate position loop and counter, see Ref. [6.2].

Numerical control systems are widely used in industry today, especially in the metal-working field. By far the most common application of NC is for metal cutting machine tools, e.g. milling, drilling, boring, turning and grinding. In addition to metal machining, numerical control has been applied to a variety of other operations, e.g. welding machines, tube bending, flame cutting, plasma arc-cutting, automatic riveting, wire-wrap machines and laser beam processes ([6.4] and [6.5]).

6.3 NC PART-PROGRAMMING LANGUAGES

Generally, an NC programming language consists of a software package comprising all the data required to produce the part, the calculation of the tool path along which the machining operations will be performed, and the arrangement of those given and calculated data in the standard format, which could be decoded to an acceptable form for a particular machine controller unit. The vocabulary words are typically English-like to make the NC language easy to use.

The best known and most comprehensive one is the Automatically Programmed Tools (APT) System, which facilitates the work of the part-program [6.3]. It consists of a series of instructions which are punched on cards or tapes and are used as an input to the computer. The APT language provides the same flexibility of expression to part-programmers that other standard programming languages provide to computer programmers. Other known computer programming systems are:

ADAPT
SPLIT
EXAPT
AUTOSPOT
COMPACT II

and many others which have been developed by various companies and can be examined in the appropriate literature.

6.4 NC PROGRAMMING WITH INTERACTIVE GRAPHICS

In Chapters 3 and 5, we dealt with the many ways in which computer-aided design and computer-aided analysis can increase the productivity of a company. Another related important reason for using a computer-aided engineering system is that it offers the opportunity to develop the data base needed to manufacture the product. In conventional engineering design, engineering drawings are prepared by draughtsmen and then used by production/manufacturing engineers to develop processing sheets. The activities involved in designing the product are separated from the activities associated with its manufacture. The barrier between design and manufacture is both time-consuming and ineffective; it involves duplication of effort by design and production engineers. In computer-aided engineering, the barrier between design and manufacture is destroyed and a direct link is established between the different phases of design, analysis and manufacture via a unified data base, as depicted in Fig. 6.2.

The use of interactive graphics in NC part-programming is an effective example of integration of computer-aided design and computer-aided manufacturing. The programming procedure is carried out on the graphics terminal of a CAD/CAM system. Using the same geometric data which defined the part during the computer-aided design process, the designer constructs the tool path using high-level commands to the system [6.1].

In many cases, the tool path is automatically generated by the software of the CAD/CAM system. The output resulting from the procedure is a listing of the APT program or the actual cutter location file which can be processed to generate the NC punched tape.

All of the major CAD/CAM system vendors offer part-programming packages. Although the features of these packages vary between the vendors, they all operate in a similar manner.

6.5 TOOL PATH GENERATION

The CAD/CIM approach for generating NC programs usually begins with the geometric definition of the object. The data base which is created during the design and analysis stages can then be used for the manufacture. The computer-aided engineering system would typically have a tool library with the various tools used in the workshop. The designer could either select one of these available tools or create his own

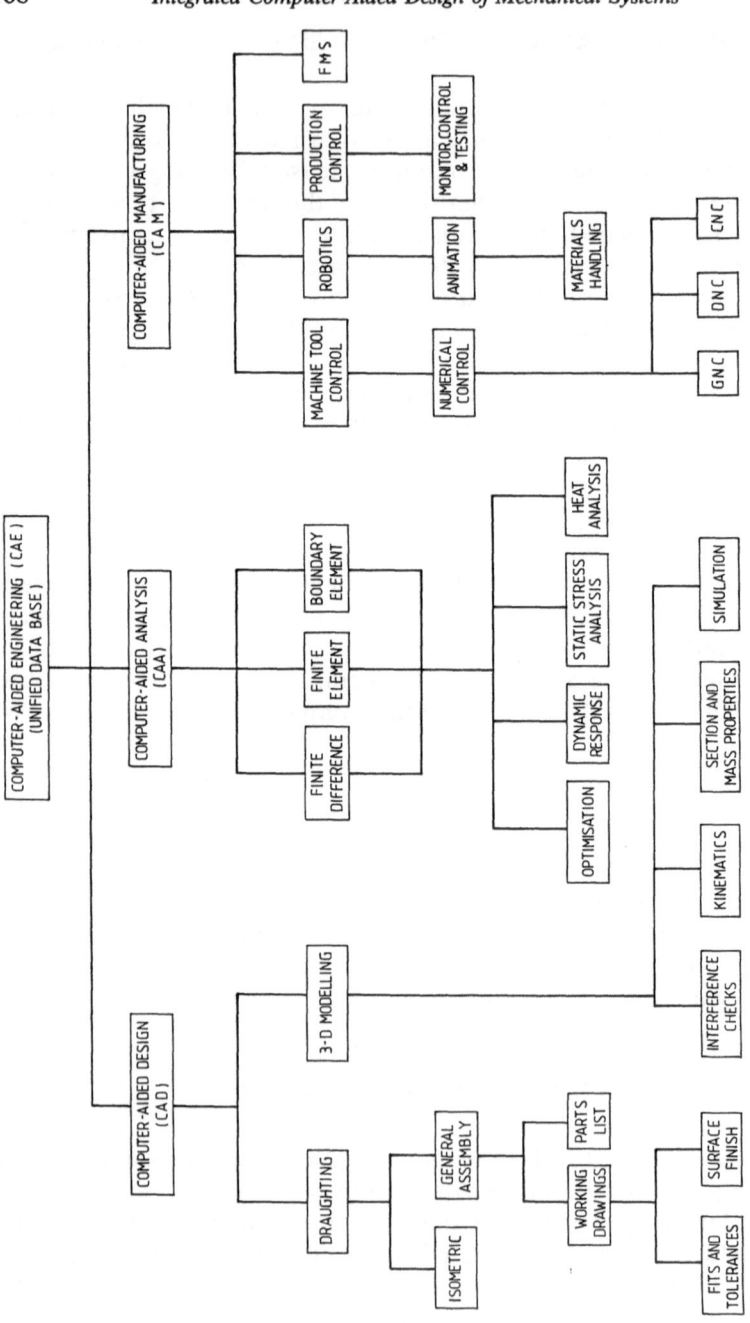

FIG. 6.2. CAD/CIM unified data base. (Abbreviations: FMS, flexible manufacturing systems; GNC, graphical numerical control; DNC, direct numerical control; CNC, computer numerical control.)

by specifying the appropriate dimensions of the cutter. This is then followed by generating the necessary tool paths. The currently available computer-aided engineering systems use an interactive approach with certain common machining routines being done automatically by the system. These include profile milling, end milling, turning, drilling and sheet metal working [6.6].

In this interactive approach, the designer generates the tool paths in a step-by-step manner with visual verification on the graphic display unit. The designer begins by defining a starting position for the cutter commanding the tool to move at prescribed speeds and feed rates along specific directions and allowing for clearance planes. The output resulting from the above procedure is a listing of automatically programmed tool (APT) paths or the actual cutter location file which can be post-processed to generate the NC punched tape, thus eliminating part-programming effort.

Let us now consider the previous case studies to illustrate the graphics approach to NC programming.

First Case Study: Computer-Integrated Manufacturing of Critical Components of Wheel-Assembly

6.6 CHOICE OF MATERIALS

Our choice of materials for the manufacture of the wheel-assembly was based upon the following:

  (i)   strength and rigidity considerations,
 (ii)   resistance to wear,
(iii)   economic considerations,
 (iv)   ease of manufacture and good machinability,
  (v)   low specific gravity for reduced centrifugal forces,
 (vi)   reduced porosity during casting.

The vast majority of the wheel parts (blades, impeller, control cage and spacers) described earlier are manufactured from a family of alloys known as 'hard irons'. These alloys are generally white cast-irons containing 15–30% chromium, 1·5–3% carbon and up to 3% molybdenum. High-chromium cast-irons are used extensively in many varied industrial applications requiring a high level of wear resistance.

Since these high-chromium alloys are cast-irons, they are inherently relatively brittle in comparison with engineering steels or even with ductile or nodular cast-irons. In those applications demanding a relatively high level of toughness, the lower carbon grades are utilised, where some resistance to abrasion is sacrificed.

The metallurgical characteristics which lead to the remarkable wear resistance of high-chromium cast-irons are hardness and micro-structure. The microstructure of white cast-irons generally consists of hard-iron carbides in a relatively soft pearlitic matrix. The influence of chromium is two-fold: firstly, the carbide phase increases further in hardness from 1200 Hv to 1600 Hv due to the presence of chromium carbides in place of iron carbides; secondly, the hardenability of the matrix phase is increased to a level such that a fully martensitic or even austenitic matrix can be achieved [6.7].

Consultation with a number of manufacturers reveals that the most suitable and versatile grade of chromium cast-irons are those containing 25% chromium; the material specification for wheel parts would typically be

25% Cr, 3% C, 0·6 Si, and 0·8% Mn

heat treated by air quenching from approximately 1000°C to a minimum hardness of 800 Hv.

6.7 MANUFACTURE OF WHEEL-ASSEMBLY

In the present design, it was thought appropriate to cast and machine impeller, blades and cage. The discs and shaft will be manufactured from rolled high strength alloy steels. The cast parts however will be manufactured by the Shell or Kröning process which utilises a fine sand coated with thermoplastic resin. This process is ideally suited for the precise definition, excellent dimensional stability and good surface finish requirement.

It is vital that any voids produced by metal contraction during cooling and solidification are excluded from the casting and retained within the feeding reservoir. This would be established by NDT-examination at the pre-production stage. In a cruder test, the casting is broken or split open to examine internal soundness.

The significance of casting integrity is of vital importance to the life of the wheel-assembly, where the smallest internal defect when exposed is rapidly scoured out by the action of the media. The same comments apply to surface defects and in this case it is highly desirable to produce good surface finish with complete freedom from visible surface defects. Dimensional tolerances are also important to ensure that maintenance and replacement can be carried out quickly and effectively.

Fixing has to be given special attention, since the abrasion-resistant nature of high-chromium irons precludes drilling and tapping. This is overcome by using cast-in steel inserts and cast-in threaded studs.

6.8 DEVELOPMENTS OF NC MACHINING PROGRAMS

The geometric definitions of the shaft and discs of the wheel-assembly were transferred from SDRC-GEOMOD to McAuto Unigraphics GMACH manufacturing system through the appropriate Initial Graphic Exchange Specification (IGES) files to develop NC machining programs for them [6.8]. Once transferred, the following must be specified in order that tool path data can be generated:

    (i) type of CNC machine tool and controller,
    (ii) the geometry of the blank material necessary for the manufacture of the component,
    (iii) the geometry of the cutter,
    (iv) position of cutter relative to component and appropriate clearance planes so that the tool can move freely, unhindered by clamps, fixtures, etc.,
    (v) maximum allowable depth of cut,
    (vi) optimum cutting speed and feed, and
    (vii) jigs and fixtures.

6.8.1 Machining of main shaft

In view of its symmetry about the axis of rotation, only one-half of the shaft was used to demonstrate the basic principles adopted in its manufacture. A piece of stock 64 mm in diameter and 415 mm in length was utilised. The maximum diameter of the finished shaft is 56 mm, thus allowing for a maximum depth of roughing cut of 3·75 mm and a maximum depth of finishing cut of 0·25 mm.

In order to demonstrate the machining program, two tool sizes will be used; one for the general roughing cut with a tip radius of 3 mm, while the other for the finishing cut, the circlip-grooves and the screwthread undercut with a tip radius of 2·5 mm. The tool changing position defined in Fig. 6.3(a) through to Fig. 6.3(c) is the point where the tool

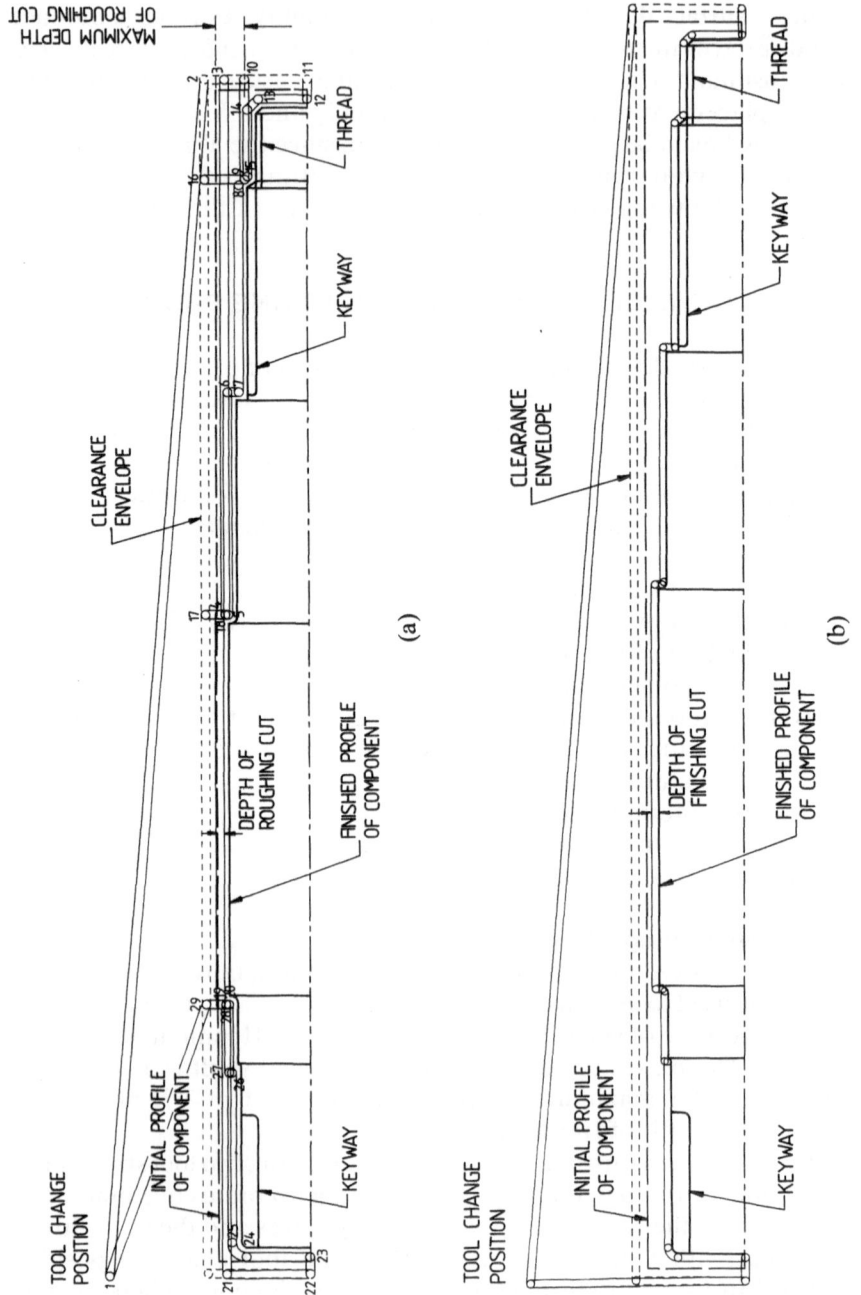

(a)

(b)

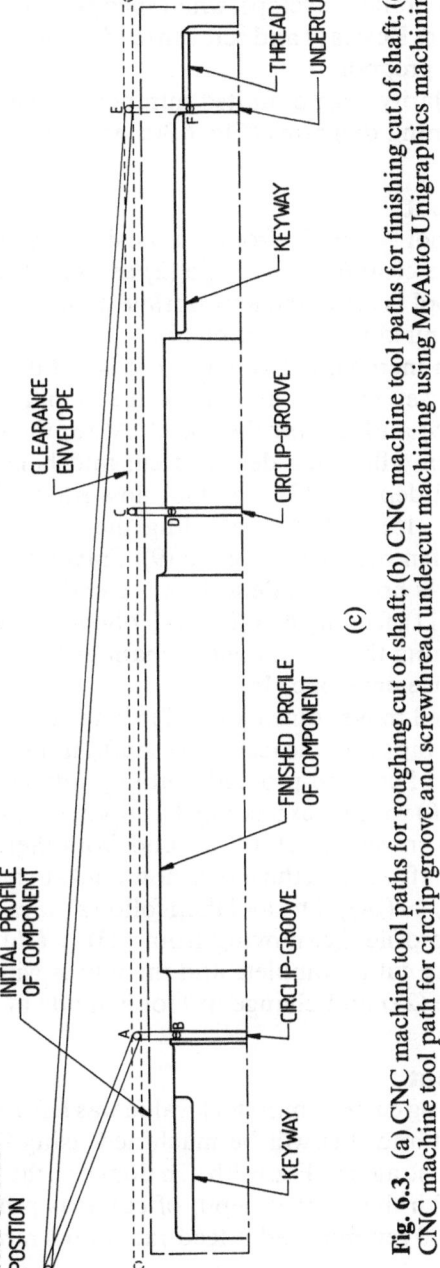

Fig. 6.3. (a) CNC machine tool paths for roughing cut of shaft; (b) CNC machine tool paths for finishing cut of shaft; (c) CNC machine tool path for circlip-groove and screwthread undercut machining using McAuto-Unigraphics machining system.

returns to, via the clearance envelope, when specific cutting instructions have been completed. Keyways and screwthread are not included in the present machining program.

For the sake of clarity, only a limited number of passes is shown on the relevant machining diagrams (Fig. 6.3(a) to (c)).

6.8.2 Roughing cut

In this case, the tool proceeds from the position marked (1) to the extreme edge of the clearance envelope (2) as depicted in Fig. 6.3(a). At this stage, the tool is set to the appropriate cutting depth at (3) and machining is continued until reaching (4).

Since the chosen incremental depth of cut is not deep enough, it is now necessary to set the remainder of the depth of cut to that corresponding to (5) and reverse the travel. At (6) the depth of cut is below the maximum allowable depth of cut and hence the tool can follow the assigned depth to (7). The tool now proceeds to (8) and it cannot proceed to (15) without exceeding the maximum allowable depth of cut. It is then permitted to go to (9). Traverse cut to clearance envelope at (10). It is now possible to face the ends of the shaft off by moving the tool to (11), feeding it to (12) and following the profile at (13) and (14) and finishing the uncompleted chamfer from (15) to (9) and back to the clearance envelope (16).

From (16), the tool moves directly to (17) via the clearance envelope, and starts the roughing cut of the rest of the shaft. At (18) the tool reaches the required depth of cut and proceeds cutting until (19). From (19) it can only move to (20) without exceeding the stipulated maximum depth of cut. From (20) it traverses cut to (21), and from there it can profile cut the rest of the shaft. At (22), the tool is now ready to face off the other shaft end (from (23) → (24)). The tool then follows the radius to (25) and then finishes the profile, i.e. moving from (25) → (26) → (27) → (28). At (28) the roughing cut is complete and the tool is withdrawn via the clearance envelope (29) and change of tool is made at (1).

6.8.3 Finishing cut

In the above roughing cut 0·25 mm stock value was left around the whole profile of the shaft and this will be machined using the appropriate finishing tool as illustrated in Fig. 6.3(b). In this case, the tool follows the assigned profile with the desired depth of cut. It is proposed to use a spindle speed of 600 rev/min and a feed rate of 0·8 mm/rev.

6.8.4 Circlips and undercut machining

In this application, the finishing tool was used to machine the two circlip-grooves and the screwthread undercut as indicated in Fig. 6.3(c). In this case, the tool is directed to move to position (A) and is instructed to cut the first circlip-groove. Upon completion, it is then allowed to go back to (A), moving along the clearance envelope to (C) at which the second circlip-groove is made. The process is then repeated for the undercut.

6.8.5 Milling cut of slots in discs

The eight blade retainer and blade guidance slots shown in Fig. 6.4 were machined from a disc with a central hole. The roughing cut which utilises an end-mill cutter of the appropriate dimensions, set to the desired depth of cut, proceeds from the tool changing position to the inner radius of the disc. Four passes are used to rough machine one complete slot to a depth of approximately 3·75 mm. The end-mill is then instructed to machine the blade retainer slot at an added depth of cut of 3·75 mm. At this stage, the semi-circular profile shown in the diagram is also machined. The milling cut is then repeated for the other slots by

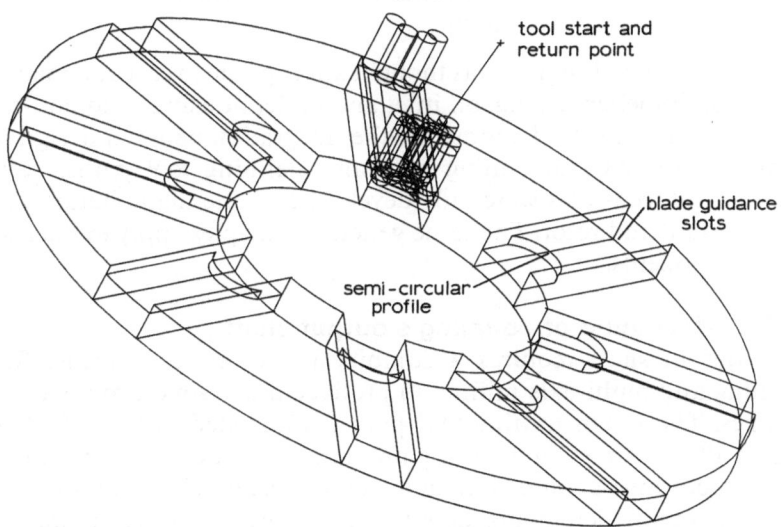

FIG. 6.4. CNC machine tool paths for milling slot in a disc using McAuto–Unigraphics machining system.

rotating the disc about its centre by 45°. For clarity, only the tool path required to form one slot in the disc is shown in Fig. 6.4.

The above rough milling operations are then followed by the finishing cut of the slots (with a maximum depth of cut of about 0·25 mm). In this case, a similar approach to that adopted in the roughing cut is proposed.

Second Case Study: Computer-Integrated Manufacturing of Fluid Coupling

6.9 CHOICE OF MATERIALS AND NC MACHINING PROGRAMS

The following materials are proposed for the critical components of the coupling:

Casing   : aluminium (LM 25) chill cast [6.9],
Impeller and rotor : aluminium (LM 25) chill cast,
Shaft   : medium carbon steel designation 080M40 according to BS 970 (formerly known as En8) [6.10].

In the present design, it was thought appropriate to cast and machine the rotor, impeller, casing, oil inlet and outlet housings and end-caps. However, in order to demonstrate the integration between design and manufacture, NC machining programs defining tool paths, cutting speeds and feed-rates were only developed for the manufacture of the output shaft and rotor. The same general principles apply to the rest of the components.

6.9.1 Machining of coupling's output shaft

The output shaft requires a combination of rough turning, finish turning and multi-axis milling to produce the desired profile and the splines. The rough turning tool path is illustrated in Fig. 6.5(a), and again all the current machining programs have been generated in the manufacturing operations module of the McAuto Unigraphics system. The maximum incremental depth of cut in the actual program was set at 3 mm for the roughing cut with a spindle speed of 400 rev/min and a feed-rate of 0·8 mm/rev, but for the benefit of improved visualisation the maximum incremental depth of cut has been increased to 7 mm. Within the rough turning program, there is also provision to leave an amount of

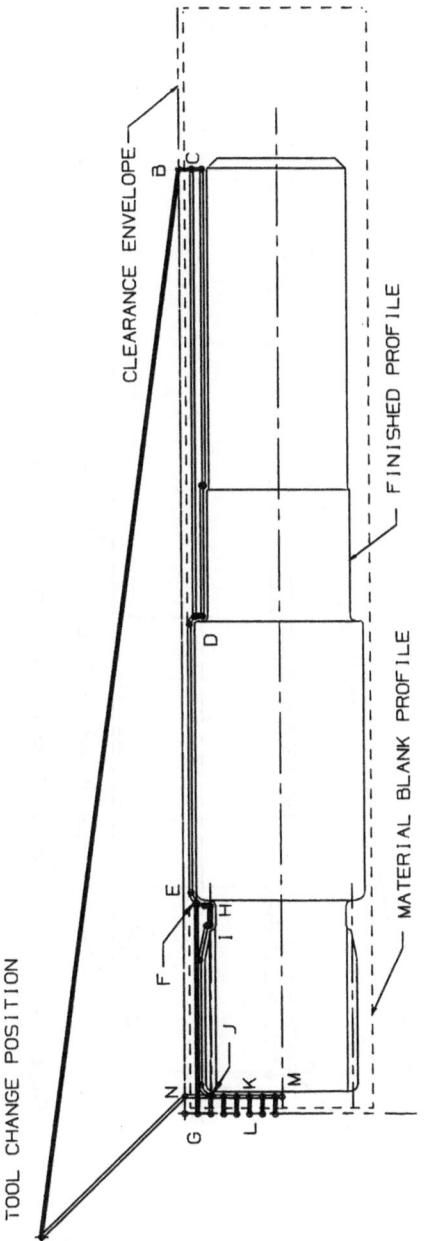

FIG. 6.5. (a) CNC machine tool paths for roughing cut of coupling's output shaft.

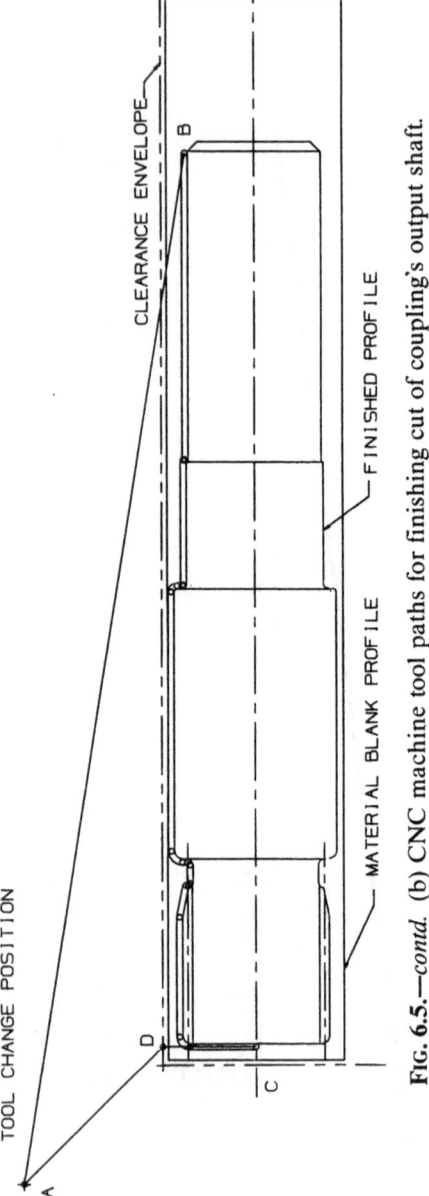

Fig. 6.5.—*contd.* (b) CNC machine tool paths for finishing cut of coupling's output shaft.

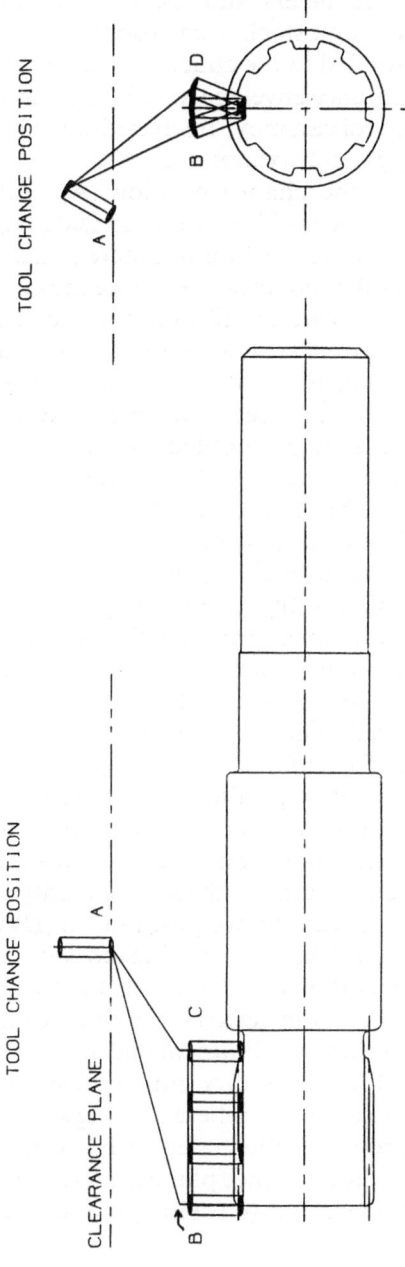

FIG. 6.5.—*contd.* (c) CNC machine tool paths for cutting of splines.

stock on the different diameters and faces for a subsequent finish turning operation. The blank stock is cut about 100 mm overlength to enable the shaft to be rotated in the chuck of a lathe. Furthermore, it is necessary to define a clearance envelope, which consists of perpendicular axes, outside which the tool can move freely without danger of colliding with chucks, centres, jigs and fixtures, etc.

The tool moves from the change position (A) to the edge of the clearance envelope (B) at a rapid feed-rate of 200 mm/min to ensure minimum wastage of time. The maximum incremental depth of cut is then indexed at (C) and the tool traverses along the component to (D). This procedure is then repeated until only the prescribed amount of stock remains on the appropriate diameters. The tool then proceeds to perform a similar operation on the other sections of the shaft. A series of similar passes enables the undercut to be machined from (F) to (H) to (I), and the splines can now be profiled as far as (J). The tool then machines the left hand face of the shaft in a series of cuts, such as (L) to (K), and when it reaches (M), the centre of the shaft, it traverses to the clearance envelope (N) and returns back to the tool change position (A) with rapid feed. This completes the roughing cut.

The finishing cut, which is illustrated in Fig. 6.5(b), is performed on the same machine, but with an appropriate finishing tool. This tool has a suitable tip radius (possibly 1 mm) to achieve tight tolerances and a smooth surface finish. A spindle speed of 600 rev/min and a feed-rate of 0·8 mm/rev are proposed. The tool proceeds from the tool change position (A), with rapid feed, and begins cutting the remaining stock at (B). It is then instructed to profile the geometry of the finished component in a single pass. In the actual program the amount of stock remaining on the different diameters and faces was 0·25 mm, but this value has been increased to 1 mm for the sake of clarity. On reaching (C) the tool can then return to the change position via (D) with rapid feed. On completion of this finishing cut, a parting tool would then be used to cut the shaft to length as this now completes the turning operations.

The profiled shaft is then transferred to a multi-axis milling machine for the cutting of the splines. A cylindrical end-mill 10 mm in diameter with a corner radius of 1 mm is used. A spindle speed of 4500 rev/min is specified to produce the correct peripheral cutting speed with a feed-rate of 10 mm/min. This process is illustrated in Fig. 6.5(c).

The tool moves from the clearance plane at (A) to begin cutting at (B) and proceeds to the undercut at (C). This process is repeated several times at varying angles to achieve the required curved profile as depicted in the figure (from (B) to (D)). The tool then returns to the change position whilst the next spline is indexed, and the milling cycle is then repeated. These three machining programs now complete the production of the output shaft.

6.9.2 Finishing cut of rotor

As explained earlier, it is envisaged that the rotor, impeller, casing, oil inlet and outlet housings and end-caps will be cast. In this case, it is assumed that the castings will be produced to a sufficient degree of geometrical accuracy to enable a single finishing operation to suffice. If a sufficient quantity of these couplings was required, it would be feasible to die-cast these components, thus producing accurate castings with good surface finish. For prototype or small batch production, however, it would not be possible to justify such expenditure on tooling, and in this case sand castings would be more suitable. Careful sand casting could still provide us with components of sufficient accuracy to warrant only a single finishing cut.

In order to demonstrate the approach adopted in developing CNC machine tool paths for intricate geometries, let us focus our attention on the machining operations necessary for the manufacture of a cast rotor.

The machining of the rotor comprises four distinct operations, as detailed below:

(i) A finish turning operation is initially performed on one end of the rotor, as shown in Fig. 6.6(a). This is necessary for the correct positioning of the bearing, the mechanical seals and for thread cutting. The rotation of the component is accomplished by clamping the flange face to a face-plate. Due to the relatively large diameters encountered in the rotor, a spindle speed of about 100 rev/min is proposed, with a feed-rate of 1 mm/rev. The appropriate cutting tool simply proceeds from the change position (A) with a rapid feed, and commences its cutting action at (B). It then profiles this end of the rotor in one pass moving as far as (C) and subsequently returning to the change position (A) via the clearance envelope at (D).

(ii) A multi-pass threading operation is required to accept the bearing lock-nut. A spindle speed of 400 rev/min is required to achieve the desired cutting speed and a feed-rate of 800 mm/ min is used to produce the appropriate pitch. A V-shaped tool is used to cut the thread in the desired end of the rotor, as detailed in Fig. 6.6(b). This operation can then be followed by a chaser to improve the surface finish and guarantee the thread pitch to a high degree of accuracy.

(iii) The rotor is then remounted in the same machine (Fig. 6.6(c)) and is rotated through 180° to enable the flange face and splines area to be machined. A finish turning tool proceeds from (A) → (B) → (C) → (D) and back to the clearance envelope at (E) and then directly to the change position (A).

(iv) The rotor is then transferred to an NC drilling machine for the final operation which involves the drilling of the clamping holes in the flange. Sixteen holes are to be drilled to a tight tolerance, thus enabling the correct mounting of rotor and casing. A series of avoidance parameters (clearance planes) are set up during the generation of the program to define a global retract position and a local retract position, where the tool can traverse freely between hole locations. A 12 mm diameter drill-bit is used with a spindle speed of 4000 rev/min and a feed-rate of 10 mm/min. The operation is shown in two different views to aid visualisation. From the change position (A), the drill proceeds to the first hole location at (B). The drilling operation is clearly explained in the local retract plane.

The drilling process also includes a facility to break the chip at pre-set intervals, hence the drill proceeds to (t) in increments of 3 mm as shown in Fig. 6.6(d)). On completion of this hole, the drill then retracts to the local retract plane (r), and then follows a similar procedure at hole locations (C) to (Q). The drill will then return to the change position, thus completing the drilling of the rotor. This is then followed by a reaming operation (not shown).

In view of the possible fatigue failure at the root of the blades of both impeller and rotor and the highly stressed regions of the shaft, a controlled shot-peening treatment has been proposed [6.11]. It is proposed to shot-peen the roots of the blades of both impeller and rotor using glass-bead media of size 150–250 μm at intensity 8 N, and the shaft using cast-steel shot S170 at intensity 12 A. Bearing surfaces will be

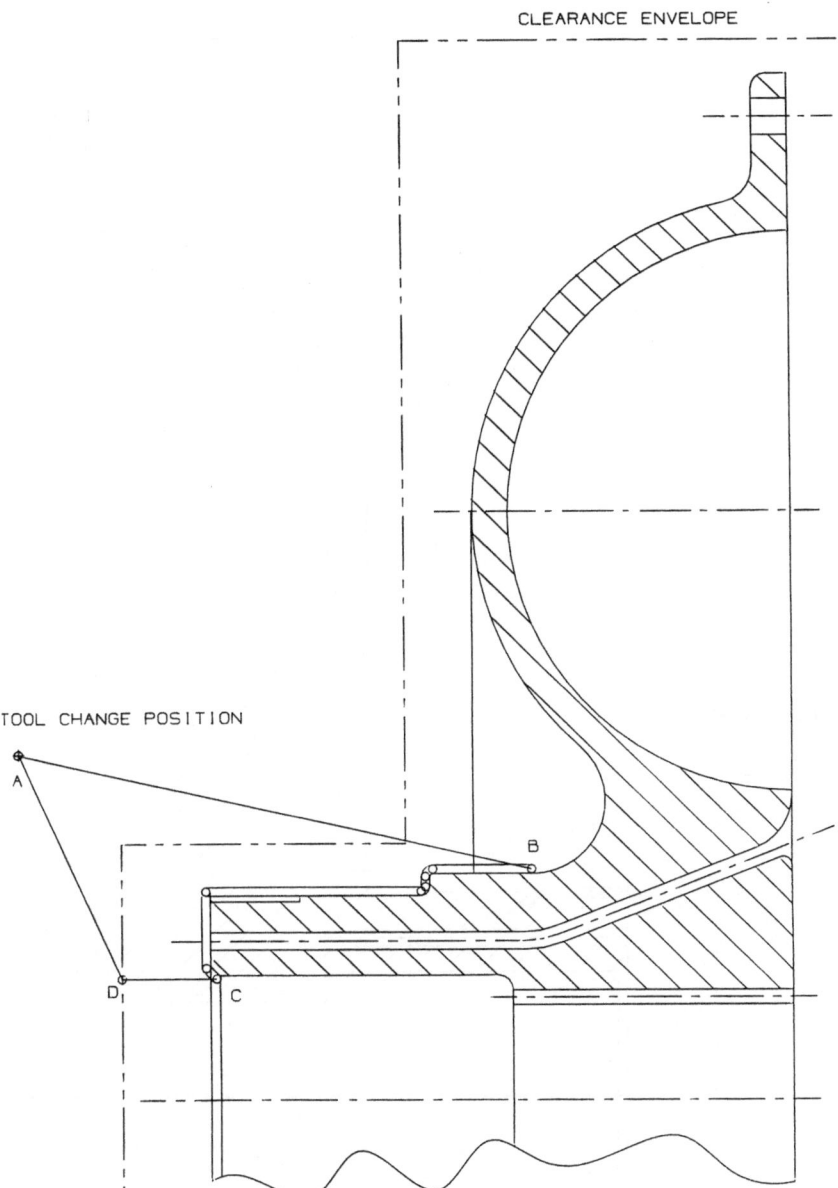

FIG. 6.6. (a) First finish turning operation performed on one end of rotor.

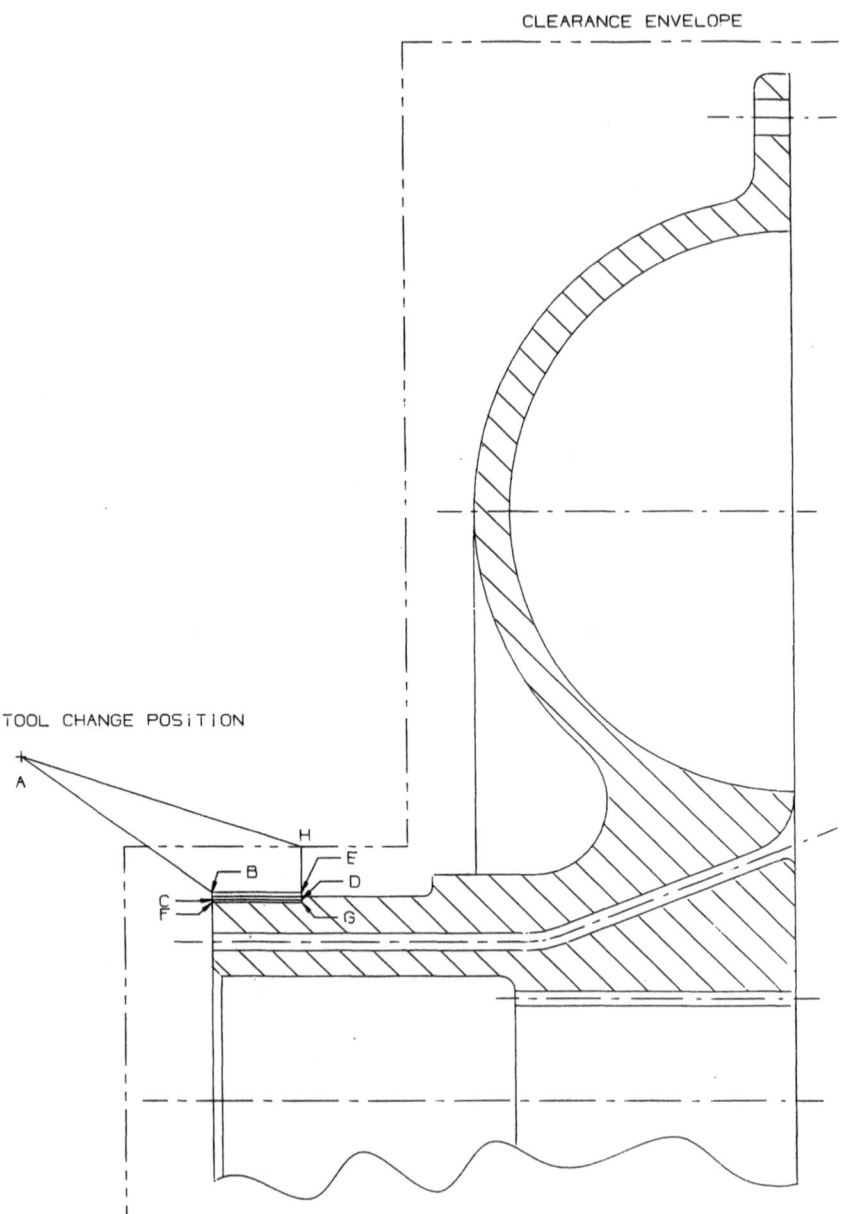

CLEARANCE ENVELOPE

TOOL CHANGE POSITION

A

H

E

B

D

C

F

G

FIG. 6.6.—*contd.* (b) Thread cutting of one end of rotor.

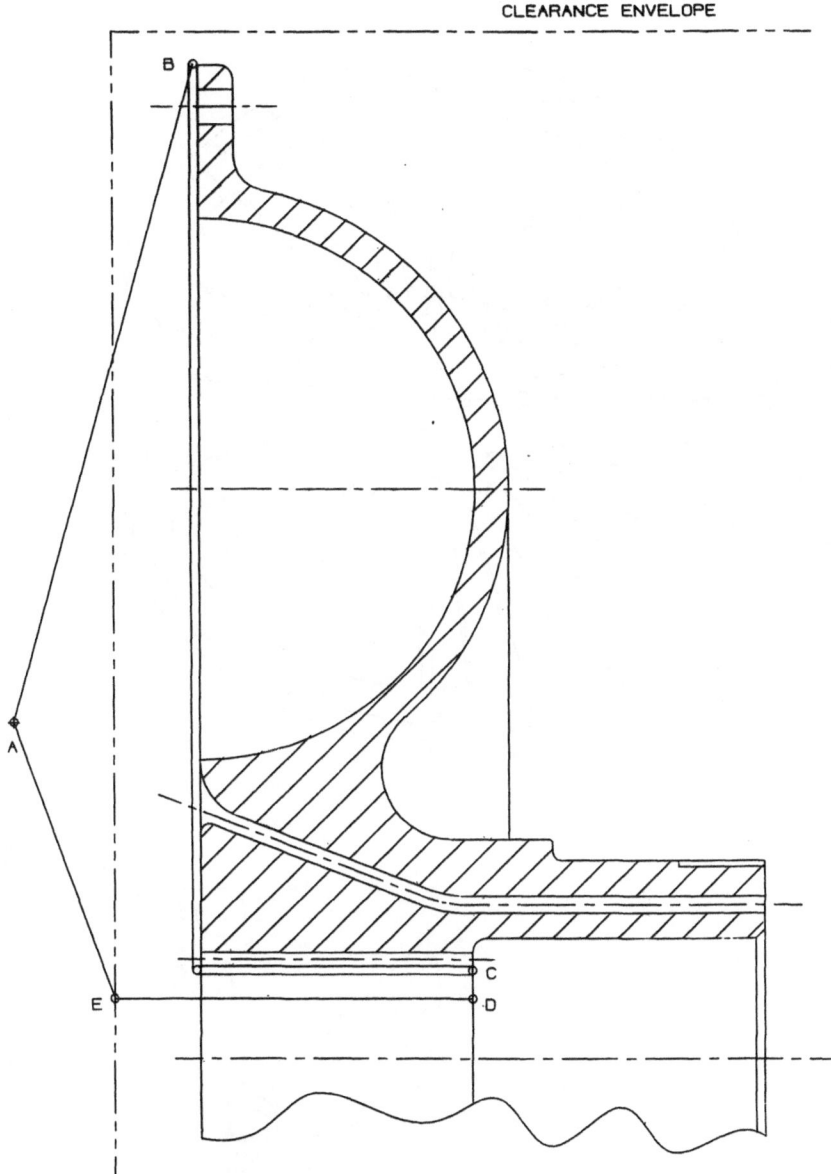

CLEARANCE ENVELOPE

FIG. 6.6.—*contd.* (c) Second finish turning operation performed on flange face and bearing area.

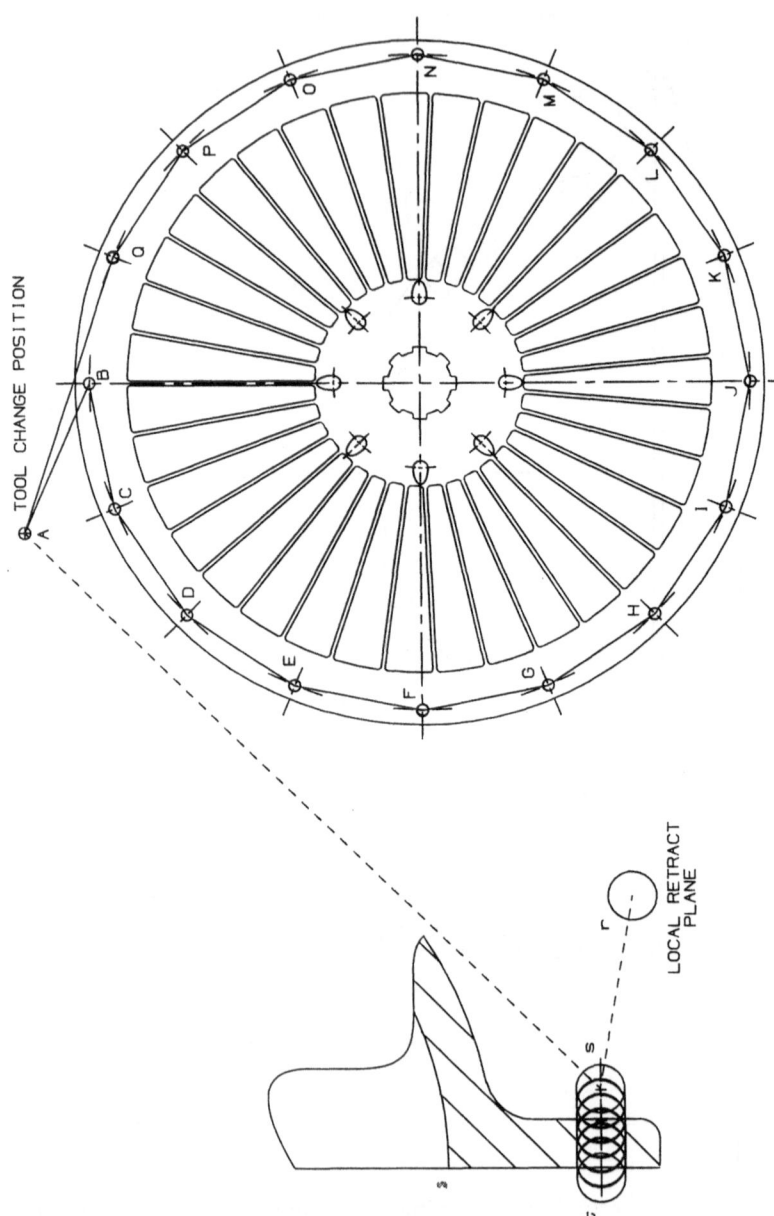

FIG. 6.6.—*contd.* (d) Drilling of holes in flange of rotor.

masked during the peening treatment. Details regarding the development of CNC machine tool paths for other critical components of the coupling can be found in Refs [6.12] and [6.13].

6.9.3 Cutter Location Source Files (CLSF)

During the generation of the tool paths, a series of cutter location source files are created. These files define cutter movement by co-ordinates and specify feed-rates, spindle speeds, etc., such that punched tapes can be produced to enable the ultimate manufacture of the product on suitably equipped machines. They are also used for documentation purposes.

Figure 6.7 gives a typical example of a source file developed by the McAuto manufacturing module for the first finishing cut of the rotor of the fluid coupling (Fig. 6.6(a)). In this file, the first statement describes the file type as being tool paths; it also initialises the reference point for tool changing position, and finally identifies the file as being (FINISH 1). The second statement is concerned with tool co-ordinate and absolute co-ordinate systems. In that statement, MSYS identifies the machine co-ordinate system and it is usually generated when a CLSF is created. Its fields are used internally by the system to relate the co-ordinates of the tool path to the absolute co-ordinate system. The first three fields represent a vector from the model space origin to the origin of the working co-ordinate system. The next six fields map the motion points described in the CLSF from the working co-ordinate system to the absolute co-ordinate system.

The statement PAINT aids the graphic verification of the tool movement, while GOTO defines the three-dimensional co-ordinates of a point in the working co-ordinate system. FEDRAT/200.00 indicates that the feed-rate used to move the tool from the change position to the edge of the clearance plane (or vice versa) is 200 mm/min. CIRCLE indicates that a curved path is being followed by the tool. In the CIRCLE statement, the following quantities are provided: the three co-ordinates of the centre of curvature; the curve axis vector; radius of curvature; total tolerance; fraction of tolerance; tool diameter; and the corner tip radius of the tool. Finally, the presence of SIL in the PAINT/PATH statement would allow the display of the tool tip radius, thus enabling the designer to check if the fillets of his/her component are accessible with the chosen tool. This dimension is given in percentage of the tip radius of tool, e.g. 100% implies that the full tip radius would appear during the graphical simulation of the machining operation.

```
TOOL PATH/1.5000,0.0000,0.0000,FINISH1
MSYS/24.62738,-196.90854,0.00000,1.00000,0.00000,0.00000,1.00000,0.00000
PAINT/TOOL
FEDRAT/200.0000
PAINT/COLOR,3
PAINT/PATH
GOTO/-286.8813,315.4085,0.0000
FEDRAT/MMPR,0.8000
GOTO/-112.8813,275.9085,0.0000
PAINT/PATH,SIL,100.0000
GOTO/-146.6274,275.9085,0.0000
CIRCLE/-146.6274,273.4085,0.0000,0.0000,0.0000,-1.0000,2.5000,0.0508,0.5000,3.0000,0.0000
GOTO/-149.1274,273.4085,0.0000
GOTO/-149.1274,269.9085,0.0000
CIRCLE/-150.6274,269.9085,0.0000,0.0000,0.0000,1.0000,1.5000,0.0508,0.5000,3.0000,0.0000
GOTO/-150.6274,268.4085,0.0000
GOTO/-223.1274,268.4085,0.0000
GOTO/-223.1274,242.0384,0.0000
GOTO/-219.5553,238.4663,0.0000
FEDRAT/200.0000
PAINT/PATH
GOTO/-251.5458,238.4663,0.0000
GOTO/-286.7196,315.6141,0.0000
PAINT/TOOL,NOMORE
END-OF-PATH
```

FIG. 6.7. A typical printout of a Cutter Location Source File for the first finish turning operation performed on one end of rotor (see Fig. 6.6(a) for related graphical representation).

6.10 CONCLUDING REMARKS

The present work illustrates clearly the use of the different aspects of computer-aided engineering (CAE) in the design of a wheel-assembly for a centrifugal peening equipment and a constant-fill fluid coupling for an auxiliary boiler feed pump application. These aspects include both two- and three-dimensional interactive computer graphics using wire-frame and solid modelling facilities, computer-aided analysis using the finite element method and finally computer integrated manufacturing. All 3-D models of the different components which make up the assemblies have been developed using SDRC-GEOMOD solid modelling facility. The detailed components were then assembled, in the system assembly section of the package, and interference checks were made. All 2-D engineering drawings of the components and the general assembly were made using McAuto–Unigraphics System.

It is clear from the current studies that interactive modelling of engineering systems allows the designer to conveniently create, manipulate, store, assemble and retrieve very complex geometries from the appropriate data base. The ability to assemble the different components of a mechanical system would ensure that design errors, which could be catastrophic at the machining or assembly stage of the manufacture, can be pre-empted by simulation of the assembly.

The present design studies also demonstrate that the different aspects of CAE represent a continuum of activity rather than a selection of separate functions. This was clearly illustrated by the use of a common data base for interactive modelling, finite element analysis, draughting and manufacture of the different components.

It is evident from the current design activities that the efficient use of computers can greatly reduce the time taken between the conception of a design to its ultimate manufacture.

In view of the complexity of the geometrical features of many of the engineering components considered and the applied loads, it seemed prudent to undertake finite element studies. Both structural stress analysis and dynamic response studies were performed to realistically evaluate the structural integrity of the different critical components and assemblies considered in this book. It is believed that the continued development of computational capabilities would allow the designer to solve succeedingly larger and computationally more demanding design problems. The capability to effectively perform more accurate analyses would allow the designer to consider structural systems which, in view of their complexity or intended function, could only be examined by sophisticated analysis.

Computer-integrated manufacturing (CIM) was performed on some of the critical components using the McAuto–GMACA software. Turning, drilling and milling operations were successfully utilised in the manufacture of the critical components of the considered mechanical systems, thus demonstrating the integration between design, analysis and manufacture.

Finally, it must be pointed out that the present book examines the detailed design, analysis and manufacturing principles of various mechanical components which are not peculiar to the present case studies but rather common to many fields of mechanical engineering. It is hoped that the material covered in this text will provide the reader with an insight into the detailed practical application of some of the essential topics of computer-aided engineering.

REFERENCES

[6.1]   M. P. Groover, *Automation of Production Systems, and Computer-Aided Manufacturing,* Prentice Hall Inc., Englewood Cliffs, NJ, 1980.

[6.2]   R. S. Pressman and J. E. Williams, *Numerical Control and Computer Aided Manufacturing,* John Wiley & Son, New York, 1977.

[6.3]   A. D. Roberts and R. C. Prentice, *Programming for Numerical Control Machines,* McGraw Hill, New York, 1978.

[6.4]   Y. Koren, *Computer Control of Manufacturing Systems,* McGraw Hill, New York, 1983.

[6.5]   R. L. Simon, CAD/CAM the foundation of manufacture automation, *Proc. Conference on CAD/CAM Technology in Mechanical Engineering,* 24–26 March, Massachusetts Institute of Technology, Cambridge, Mass., pp. 221–35, 1982.

[6.6]   Unigraphics Technical Manual, GMACH Users Operational Description, CAD/CAM Service, McDonnel Douglas Automation Company, McDonnel Douglas Corporation, USA, 1983.

[6.7]   R. W. Durman, Observations on the use of hard-metal castings in the shot-blasting industry, *J. Mech. Working Technology,* 8, pp. 217–23, 1983.

[6.8]   P. F. Abrahams, Computer-Aided Engineering of Centrifugal Peening Equipment, MSc Thesis, Cranfield Institute of Technology, 1985.

[6.9]   BSI 490, Aluminium and Aluminium Alloy Ingots and Castings, UDC 669.72-412-14, British Standards Institution, 1970.

[6.10] J. H. E. Fox, *An Introduction to Steel Selection: Part 1,* Oxford University Press, 1979.

[6.11] S. A. Meguid (Editor), *Proc. Second International Conference on Impact Treatment Processes,* 22–26 September, Cranfield, UK, 1986.

[6.12]   P. A. Thomson, ICAD-CAM of a Variable Speed Drive for a Crude Oil Pump, MSc Thesis, Cranfield Institute of Technology, 1986.

[6.13]   P. K. Prakash, Computer-Aided Analysis of a Variable Speed Drive for a Crude Oil Pump, MSc Thesis, Cranfield Institute of Technology, 1986.

Index

Index